Nuclear Engineering Division 2017

Core Programming Area at the 2017 AIChE Annual Meeting

Minneapolis, Minnesota, USA
29 October - 3 November 2017

ISBN: 978-1-5108-5801-5

Printed from e-media with permission by:

Curran Associates, Inc.
57 Morehouse Lane
Red Hook, NY 12571

Some format issues inherent in the e-media version may also appear in this print version.

Printed by Curran Associates, Inc. (2018)

AIChE
120 Wall Street, FL 23
New York, NY 10005-4020

Phone: (800) 242-4363
Fax: (203) 775-5177

www.aiche.org

Additional copies of this publication are available from:

Curran Associates, Inc.
57 Morehouse Lane
Red Hook, NY 12571 USA
Phone: 845-758-0400
Fax: 845-758-2633
Email: curran@proceedings.com
Web: www.proceedings.com

TABLE OF CONTENTS

Some abstracts in this book refer to images that were not made available by the publisher

503670 Digital Alchemy for Assembly Engineeringc

Abstract:

Back in the day, Stuart Churchill was a pioneer in the use of computers for modeling chemical and thermal processes. Since that time, the role of computing in chemical engineering has become indispensable. In this talk, we present new computational methods we call "digital alchemy" that allow for predictive modeling of thermodynamic assembly, and permit the design of building blocks optimized for assembly into target structures with desired properties. In this approach, standard thermodynamic ensembles are generalized to include alchemical variables and their thermodynamic conjugates, allowing for rapid, real-time computational investigation of large, complex design spaces.

503671 Improved Algebraic, Numerical, and Graphical Representations in Fluid Mechanics

Abstract:

The concept of equivalents is applied to fully developed flows through straight channels, and two applications are introduced: the equivalents that produce equal rates of flow through channels of different cross-section, and the diminishment that occurs between the regimes of laminar and turbulent flow. The validity of the eddy diffusivity and the invalidity of the mixing-length as measures of the shear stress due to turbulence have perhaps never before been demonstrated so clearly. The possible choices of primary variables and the consequences are also examined. The objective of these explorations is an improved understanding of the basic fluid mechanics by teachers, students, and industrial practitioners.

503672 The Scaling of Turbulence Near the Wall and the Churchill Turbulent Flux Correlation: Insights with Lagrangian Simulations

Abstract:

The evaluation of mean velocity and mean temperature in wall-bounded turbulent flows has been based on the "classical" theory of turbulence and the law of the wall. The viscous wall parameters, namely the friction velocity, the friction temperature, and the kinematic viscosity of the fluid have been used to scale not only the mean values of velocity, but also other mean turbulent quantities close to the wall, including the Reynolds stresses and the turbulent velocity and temperature fluctuations. Stuart Churchill, as he questioned the validity of the Reynolds analogy between momentum and heat or mass transfer in his works [1], he realized that extending the scaling of momentum to the scaling of scalar quantities in turbulence was not a sound approach. He proposed a scaling of turbulence that depends on the turbulent momentum flux and the turbulent heat flux. In his AIChE Institute Lecture (in Miami, 1998), Churchill presented a summary of his work on this topic, and closed form solutions for the mean velocity and temperature [2]. The turbulent Prandtl number, the most significant parameter for the modeling of turbulent transport not only in theory but also in modern CFD commercial software packages, was also interpreted physically based on these ideas [3]. We have used direct numerical simulation results and results from Lagrangian simulations [4] for the transport of passive scalars in turbulent channel flows to explore Churchill's ideas on turbulent flow scaling and to develop an updated theory for the scaling of heat transport in turbulent flows [5-7]. His passion to remove empiricisms and approximations from the development of predictive correlations in turbulent transport has been an inspiration. In this talk, we will revisit the Churchill scaling in view of our simulation results, and will discuss the counterintuitive finding that wall turbulence can be used to separate rather than mix substances under special circumstances [8].

1. Churchill, S. W., 1996, "A Critique of Predictive and Correlative Models for Turbulent Flow and Convection," Ind. Eng. Chem. Res., 35, pp. 3122–3140.
2. Churchill, S.W., 2000, "Progress in the Thermal Sciences: AICHE Institute Lecture," AIChE Journal, 46(9), pp. 1704-1722.
3. Churchill, S. W., 2002, "A reinterpretation of the Turbulent Prandtl Number," Ind. Eng. Chem. Res., 41, pp. 6393–6401.
4. Papavassiliou, D.V, 2002, "Turbulent transport from continuous sources at the wall of a channel," Int. J. Heat Mass Transfer, 45(17), pp. 3571-3583.
5. Le, P.M., and Papavassiliou, D.V., 2008, "On the scaling of heat transfer using thermal flux gradients for fully developed turbulent channel and Couette flows," Int. Commun. Heat Mass Transfer, 35(4), pp. 404-412.
6. Srinivasan, C., and Papavassiliou, D. V., 2011, "Prediction of Turbulent Prandtl Number in Wall Flows With Lagrangian Simulations," Ind. Eng. Chem. Res., 50(15), pp. 8881–8891.
7. Srinivasan, C., and Papavassiliou, D.V., 2013, "Heat transfer scaling for wall bounded turbulent flows," Applied Mechanics Review 65(3), Art. 031002.
8. Nguyen, Q., Srinivasan, C., and Papavassiliou, D.V., 2015, "Flow induced separation in wall turbulence", Phys Rev E, 91, Art. 033019.

503674 Flow Boiling Using a Piranha Pin Fin Heat Sink

Abstract:

Flow Boiling using a Piranha Pin Fin Heat Sink

Corey Woodcock, Xiangfei Yu, Yoav Peles[1] and Joel Plawsky

Rensselaer Polytechnic Institute; [1]University of Central Florida

The *Piranha Pin Fin* (*PPF*) microdevice is a heat exchanger engineered to dissipate extreme thermal loads using either dielectric fluids such as HFE7000 or water. Fabrication of the device is CMOS compatible and the first generations of the device have been fashioned from silicon substrates. The PPF operates on the principle that heat transfer would be more efficient if heated fluids could be removed nearly as fast as they are created. This process would keep the driving force as high as possible. Thus the heat exchanger is built as two layers; a primary substrate that dissipates the bulk of the heat and shunts the heated fluids to a booster section that continues to heat the fluid and extends the total surface area of the device. Three generations of devices were fabricated. The first two generations were single-layer systems designed to help understand the operation of the primary layer and optimize the shape and location of the pin fins. The third generation was the first, two-layer system to be fabricated and tested.

Figure 1 Representative PPF array

Correlations for the heat transfer coefficient, h_{tp}, based on the Boiling Number and the Weber Number in the PPF design was developed and validated with Gen II. The discrepancy of heat transfer coefficient between predicted and measured values was within 20%. The ratio of two-phase to single-phase pressure drop was correlated with the Boiling Number (Bo) and the Jakob Number (Ja). The discrepancy of the two-phase multiplier between prediction and experimental measurements were within 12%.

The third generation *PPF* heat sinks were the first series to provide both a primary substrate populated with *PPF*s, as well as a booster substrate populated with solid micro pin fins. The substrates are micro-aligned and fusion-bonded to provide a full three dimensional (3D) architecture with bulk silicon properties. Experimental results from the Gen III *PPF* microdevices provide a proof of concept of the true *PPF* heat sink concept with fluid venting into multiple substrates. At low flowrates (G = 670 [kg/s/m^2]) the Gen III *PPF* heat sink has been shown to dissipate heat loads approaching 1.1 [kW/cm^2] at an exit quality of 95%, while at higher flowrates (G = 3,300 [kg/s/m^2]) the thermal limit of the heat sink remains unknown due to exceeding the available power supply at >1.4 [kW/cm^2].

503673 Transport Problems in the Spirit of Stuart Churchill for Teaching and Research at the University of Michigan

Abstract:

Stuart Churchillâ€™s solutions to important heat transport problems have greatly influence both pedagogy and research. The correlation of Churchill and Chu for the natural convection heat transfer to or from a vertical plate, over an astonishing range of Rayleigh numbers is based partly on a similarity solution of a type common in heat, mass, and momentum transfer for problems in chemical engineering. One such problem is asphaltene or wax transport and deposition onto the wall of an oil pipeline, which can be solved using similar methods, as will be discussed. In addition, such problems, and the wealth of their applications make for excellent motivations and examples in the teaching of graduate or undergraduate transport, as will be discussed in the talk.

509054 Solve this! Fundamental Approach to Problem Solving in Industrial Processes I (Invited Talks)

Abstract:

Solving complex plant problems is an everyday exercise that affects all projects and budgets, and it is very rarely introduced to students at any level. In fact, by the time young engineers are few years into working in industry, most have lost the engineering skills that they developed in school. We hope to encourage future engineers to retain those skills and learn the importance of plant problem-solving.

504167 How the Nuclear Engineering Division Influenced My Career

Abstract:

The Nuclear Engineering Division (NED) has been my professional home for the last 25 years. As a young researcher working in pilots plants at the Savannah River Site, I began presenting each year at NED sessions. As my career progressed, I also got involved in chairing NED sessions. After my kids got into high school, I had more time to support my profession through AIChE and additionally got involved in leadership in NED, serving as chair in 2008. This was a springboard to help me get involved in other AIChE activities such as participation and leadership in the Career and Education Operating Council (chair in 2013) and being elected to the AIChE Board of Directors (2014-2016).

NED is the oldest AIChE division, established in 1954. NED may be a small division in AIChE but it has made significant impacts to our profession. The highest attendance of any AIChE webinar (presented by Stuart Arm) occurred soon after the Fukushima Nuclear disaster in Japan. The current AIChE president, Bond Calloway, got his leadership start in NED and is the 2014 recipient of the Robert E Wilson Award. A look down the list of Wilson award winners includes a number of industry giants.

My talk will focus on the people and the technology that have influenced my career. Nothing I do can be done in a vacuum so I will discuss some of the people that have impacted my career and what I have learned from them. I will share some stories about the work we did together and the impact it had on the processing of high-level radioactive waste. I am fortunate to have found work that is both rewarding to me and is of benefit to my community, my state, our nation, and even my profession. I will humbly receive an award I do not feel qualified to accept, but know that if there is any merit in my receiving it, it is due to the exceptional people who have supported me in my work.

502785 The Preparation of Titanate Nano Materials Removing Efficiently Cs-137 from Radioactive Waste Water

Abstract:

Cs-137, the radioactive fission products of uranium, can be easily dissolved in water during accident of nuclear power plant, such as Chernobyl, Three Mile Island, Fukushima accidents. The concentration of Cs in the groundwater around the nuclear power plant exceeded the standard value almost 10,000 times after the Fukushima accident. The adsorption capacity of Titanate nano-materials for radioactive cation (Cs+) is very strong. Moreover, the radioactive ion can be tightly contained in the nanotubes or nanofibers without reversible adsorption and it can safely be fixed. In addition, the nano-material has good chemical stability, thermal stability and mechanical stability to minimize the environmental impact of nuclear waste and waste volume. The preparation of titanate nanotubes or nanofibers was studied by hydrothermal methods and chemical kinetics of removal of Cs by nano-materials was obtained. The adsorption time with maximum adsorption capacity and the effects of pH, coexisting ion concentration and the optimum adsorption conditions on the removal of Cs by titanate nano-materials were also obtained. The adsorption boundary curves, adsorption isotherm and the maximum adsorption capacity of Cs-137 as tracer on the nano-materials were studied in the research. The experimental results showed that the removal rate of Cs-137 in 0.01 tons of waste water with only 1 gram nano-materials could reach above 98%, according to the optimum adsorption conditions.

492542 Aging Processes of Ag-Exchanged Mordenite and Ag-Functionalized Silica Aerogel in Spent Nuclear Fuel Reprocessing Off Gases

Abstract:

Aging processes of Ag-exchanged mordenite and Ag-functionalized silica aerogel in spent nuclear fuel reprocessing off gases

Yue Nan,[1] Carter W. Abney[2], Seungrag Choi,[1] Jiuxu Liu[1], Lawrence L. Tavlarides[1*]

[1]*Department of Biomedical and Chemical Engineering, Syracuse University, 329 Link Hall, Syracuse, NY 13244, USA*

[2]*Chemical Sciences Division, Oak Ridge National Laboratory, P.O. Box 2008, Oak Ridge, Tennessee 37831, USA.*

Email: lltavlar@syr.edu (Tavlarides); ynan@syr.edu (Nan)

The aging issue is one major problem in the removal of radioactive iodine vapors from spent nuclear fuel reprocessing off-gas streams using silver containing adsorbents. During the off-gas treatment, columns packed with adsorbents are expected to be exposed to off-gas streams containing air, iodine vapors, H_2O vapor, NO and NO_2 over extended periods of time (up to 6 months). Consequently, the iodine adsorption capacity of the adsorbents would decrease over time due to a series of chemical and physical changes to the adsorbents. Therefore, it is of importance to understand the aging processes of the silver containing adsorbents in off-gas streams for better designing of the treatment systems.

As the most outstanding adsorbents for iodine retention in the off-gas treatment, aging processes of Ag-exchanged mordenite (Ag^0Z) and Ag-functionalized silica aerogel (Ag^0-aerogel) have been studied over the past few years. Previous studies have experimentally investigated the aging effects of Ag^0Z and Ag^0-aerogel in different off-gas species including dry-air, humid-air, 2% NO_2/air, and 1% NO/air, and found that the iodine adsorption capacity of Ag^0Z and Ag^0-aerogel decreased significantly under most of the studied aging conditions. For example, Ag^0Z lost 60% of iodine adsorption capacity after aging in a humid-air flow for 4 months and lost 78% of the capacity after aging in a batch reactor filled with 1% NO/air for 2 months, while Ag^0-aerogel lost 22% and 43% of iodine capacity, respectively, under the same conditions. Also, 30% iodine capacity loss was observed for Ag^0Z after 2 months aging in a 2% NO_2/air batch, but no significant capacity loss was observed for Ag^0-aerogel. These results from previous studies indicated that both adsorbents had different capacity losses in different aging gases, and under the same aging conditions Ag^0Z and Ag^0-aerogel had different relative capacity losses. However, reasons for these differences are not clear and, therefore, the mechanisms of the aging processes need to be determined.

In this study, experimental and analytical work has been proposed and is on-going to determine the mechanisms of aging processes of Ag^0Z and Ag^0-aerogel in different aging gases. A continues-flow aging experimental system has been constructed to produce aging samples of Ag^0Z, Ag^0-aerogel and Ag foil in dry-air, humid-air, 2% NO_2/air and 1% NO/N_2 flows at 150 °C for periods of time up to 6 months. The continuous-flow aging experiments with 2% NO_2/air and 1% NO/N_2 are expected to provide more information than the batch experiments in previous studies. Iodine adsorption capacities of the aged samples are determined by iodine loading experiments. Chemical and physical changes to the aged samples are determined by analytical methods including X-ray diffraction (XRD), X-ray absorption fine structure (XAFS), pair density function (PDF), scanning electron microscopy (SEM) and Brunauer-Emmett-Teller (BET) analysis. Chemical forms of Ag (Ag^0, Ag_2O, $AgNO_2$, $AgNO_3$) and corresponding quantities in the adsorbents, as well as changes to the surface area, pore volume and size, and morphologies of the adsorbents will be determined via these analyses. The mechanisms of the aging processes are expected to be determined.

At this writing, samples of Ag^0Z, Ag^0-aerogel and Ag foil aged in dry-air and humid-air for 1 month have been obtained. Results of iodine loading experiments showed that Ag^0Z lost 10% and 45% iodine capacity after aging in dry-air and humid-air, respectively, and Ag^0-aerogel lost 13% capacity after aging in dry-air (the iodine adsorption experiment with humid-air aged sample is in progress). Characterizations of the samples by the analytical methods mentioned above is in progress. In addition, the aging experiments of Ag^0Z, Ag^0-aerogel and Ag foil in dry-air, humid-air, 2% NO_2/air, and 1% NO/N_2 flows for up to 6 months are on-going and those results will also be reported.

499045 Adsorption Models for Treatment of Nuclear Reprocessing Off-Gases

Abstract:

The off-gas stream from reprocessed nuclear fuel contains a number of radioisotopic gases including HTO (^{3}HHO), $^{129}I_2$, $^{129}ICH_3$, $^{14}CO_2$, ^{85}Kr, and ^{135}Xe. Several of these gases are either hazardous to human health or the environment, so there is a great interest in the efficient removal, recovery, and/or disposal of the gases from the off-gas stream. Effectively modeling the separation processes involved in this system will aid both scientists and engineers in the design and implementation of the adsorption systems necessary to perform these tasks. In this work, we have developed a compartmentalized modeling approach for single and multi-component adsorption equilibria and mass transfer kinetics. Each kernel in the model is separated, such that they can all work independently or in conjunction with other kernels. The equilibrium models include software to analyze experimental data and obtain sets of equilibrium parameters, as well as make predictions of the adsorbed amounts attainable under varying conditions of temperature and pressure. Additionally, we have developed a generalized routine for predicting the adsorption equilibrium behavior of a mixed gas system based on the equilibrium parameters of each individual species in the gas phase. The diffusion mass transfer models were developed specifically for spherical, biporous pellets composed of adsorbent crystals held together in a matrix of a binder material. This is the typical configuration of commercial zeolites and mordenites used for water vapor separations. In these models, we consider the mechanisms of film mass transfer, pore diffusion, and surface diffusion to dominate the uptake rates and assume local equilibrium between the gas and solid phases at the boundary of the adsorbent crystals. These models have been examined for water vapor adsorption in a commercial molecular sieve 3A-type adsorbent, and the multi-component equilibrium model has been verified against several sets of literature data. We plan to utilize these models into a macro-scale mass transport model to simulate the capture efficiency of a fixed-bed column for the separation of gas mixtures.

502334 Selective Adsorption for Kr/Xe Capture and Separation

Abstract:

Used nuclear fuel reprocessing generates volatile radionuclides that will need to be controlled to meet United States regulatory emission limits. As part of a US Department of Energy (DOE) sponsored team, Idaho National Laboratory (INL) is conducting ongoing research of physical adsorption processes as a simpler, more economical alternative to cryogenic distillation for the capture and separation of krypton and xenon from these off gas streams. Two novel engineered composite sorbents, developed, patented, and evaluated at INL, have shown the capability to capture and separate krypton and xenon from air. Sorbent evaluation results are promising for scale-up to potential process conditions, with demonstrated flexibility to operate successfully over a broad range of flow conditions and gas compositions. Current research is focused on emulating potential process conditions and desorption optimization to ensure successful operation. While this research has been directed at krypton and xenon, the sorbent production process, with relatively simple modification, could be applicable to other gas adsorption processes as well.

494983 Continuous Monitoring of Ni Corrosion in Eutectic LiCl-KCl-UCl_3 Using High Temperature Voltammetry

Abstract:

Pyrochemical reprocessing in molten chloride is a leading technology option for the reprocessing of used nuclear fuel. This process includes several high temperature (773 K or higher) unit operations, including electrorefining. Electrorefining separates U or U/TRU from the used fuel via electrodeposition on a cathode using a molten LiCl-KCl-UCl_3 electrolyte. Stainless steel is the preferred candidate material for the materials of construction in an electrorefiner in contact with the molten salt. To better understand the potential for corrosion of stainless steel in an electrorefiner, electrochemical experiments were performed with Ni in contact with LiCl-KCl-UCl_3, while varying concentrations of H_2O vapor were bubbled into the salt. The water vapor varied from 20 to 950 ppm in argon and was bubbled into the salt using a ceramic tuyere at 773 K. Electrochemical measurements such as open circuit potentiometry and cyclic voltammetry were used to monitor the reactions occurring in the molten salt in near real time and real time, respectively. Evidence of oxidation of Ni to $NiCl_2$ was observed in addition to precipitation of uranium from the salt via reaction with water to form oxides. Conclusions will be presented regarding the maximum allowable moisture concentration in the cover gas for the electrorefiner to prevent stainless steel corrosion. Consequences of accidental ingress of air into a pyrochemical reprocessing facility will also be discussed.

DEPARTMENT OF
Engineering Physics
UNIVERSITY OF WISCONSIN–MADISON

A REVIEW OF IR SPECTROSCOPY OF MOLTEN FLUORIDE SALTS

WILL DERDEYN, RALUCA SCARLAT

Department of Engineering Physics
University of Wisconsin - Madison
wderdeyn@wisc.edu

HEAT AND MASS TRANSPORT GROUP | heatandmass.ep.wisc.edu

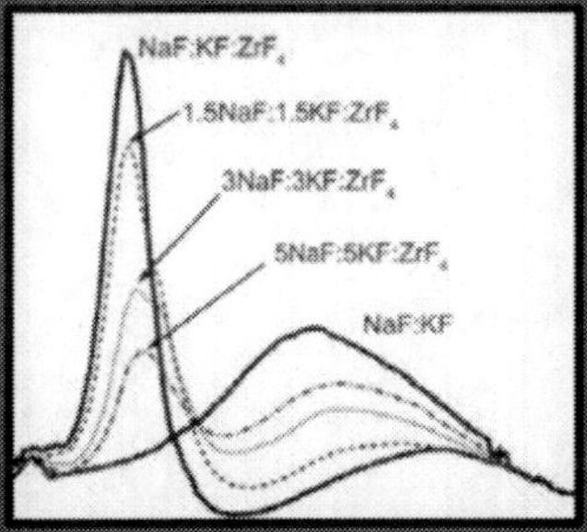

2017 AIChE Annual Meeting, 31 Oct 2017
Minneapolis Convention Center
Minneapolis, MN

[6] J. K. Wilmshurst, 'Infrared Spectra of Molten Salts', J. Chem. Phys., vol. 39, no. 10, pp. 2545-2548, (1963).
Background image courtesy of Jeff Miller/University of Wisconsin Madison

Outline

1. Motivation
2. Background Physics
3. Review of FLiBe Literature Data
4. RHT Analysis
5. Uncertainty
6. Conclusions and Path Forward

WILL DERDEYN
HEATANDMASS.EP.WISC.EDU 2

Molten Fluoride Salts: Why IR Spectroscopy?

- Molten fluoride salts are generating a lot interest in the nuclear engineering community
- We know surprisingly little about their fundamental chemical structure.
- Several salts and methods to choose from:

FLiBe (Well-studied)		Ab Initio Molecular Dynamics
FLiNaK		Neutron Scattering
KF-ZrF4		X-ray Diffraction
		Raman Spectroscopy
	Low overhead	IR Spectroscopy

- RHT is potentially significant in MSRs and FHRs due to large temperature gradients in heat exchangers, for instance in the TCHX [1], where salt exchanges heat with water
- IR absorption coefficient critical parameter needed to properly model RHT behavior

[1] C. Andreades, A. T. Cisneros, J. K. Choi, A. Y. K. Chong, M. Fratoni, S. Hong, L. R. Huddar, K. D. Huff, D. L. Krumwiede, M. R. Laufer, M. Munk, R. O. Scarlat, N. Zweibaum, E. Greenspan, and P. F. Peterson, "Technical Description of the 'Mark 1' Pebble-Bed Fluoride-Salt-Cooled High-Temperature Reactor (PB-FHR) Power Plant," University of California - Berkeley, UCBTH-14-002, 2014.

Outline

1. **Motivation**
2. **Background Physics**
3. **Review of FLiBe Literature Data**
4. **RHT Study**
5. **Uncertainty**
6. **Conclusions and Path Forward**

Background: Planck's Law

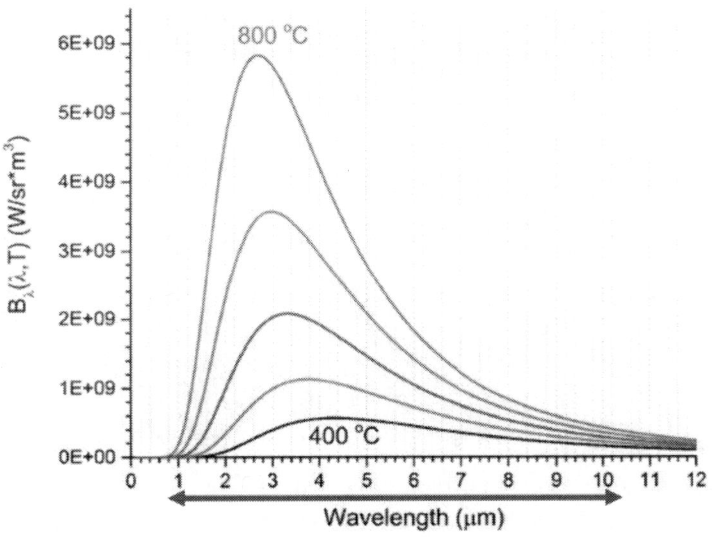

Wavelength range of interest: 1 – 10 μm

- Planck's Law describes the spectral distribution of emitted power by a blackbody at a given temperature
- We choose 400 – 800 °C as our temperature range of interest

Outline

1. Motivation
2. Background Physics
3. Review of FLiBe Literature Data
4. RHT Analysis
5. Uncertainty
6. Conclusions and Path Forward

FLiBe IR Spectrum

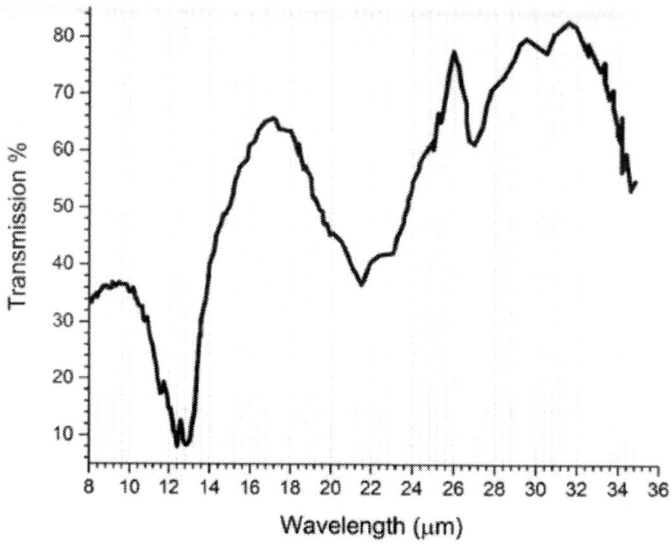

- IR transmission spectrum of FLiBe collected by Quist [3]
- Experimental description: sample ground and pressed between AgBr plates
- No path length reported

[3] A. S. Quist, J. B. Bates, et al., 'Raman spectra of tetrafluoroberyllate ion in molten sodium fluoride and lithium fluoride to 686 C', J. Phys. Chem., vol. 76, no. 1, pp. 78–83, (1972).

WILL DERDEYN
HEATANDMASS.EP.WISC.EDU
7

FLiBe Absorption

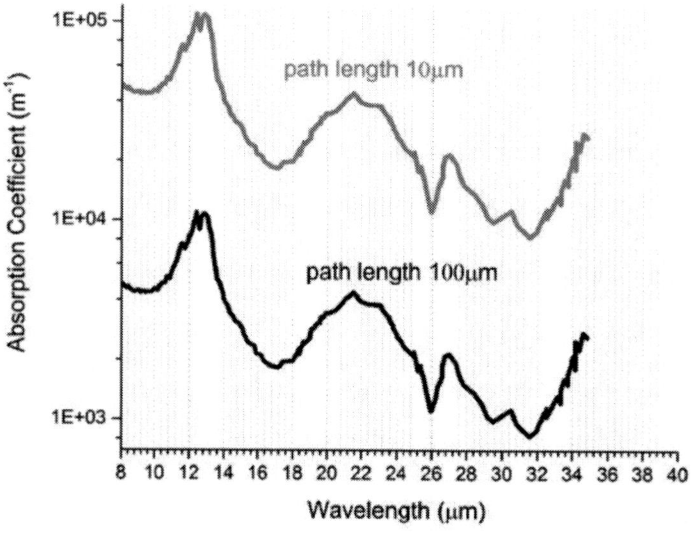

- For the sake of further analysis, we will assume the actual path length of Quist's sample lay between 10 -100 μm
- The two resultant absorption spectra will serve as bounds over an uncertainty range for FLiBe

[3] A. S. Quist, J. B. Bates, et al., 'Raman spectra of tetrafluoroberyllate ion in molten sodium fluoride and lithium fluoride to 686 C', J. Phys. Chem., vol. 76, no. 1, pp. 78–83, (1972).

WILL DERDEYN
HEATANDMASS.EP.WISC.EDU
8

LiF Absorption Spectra

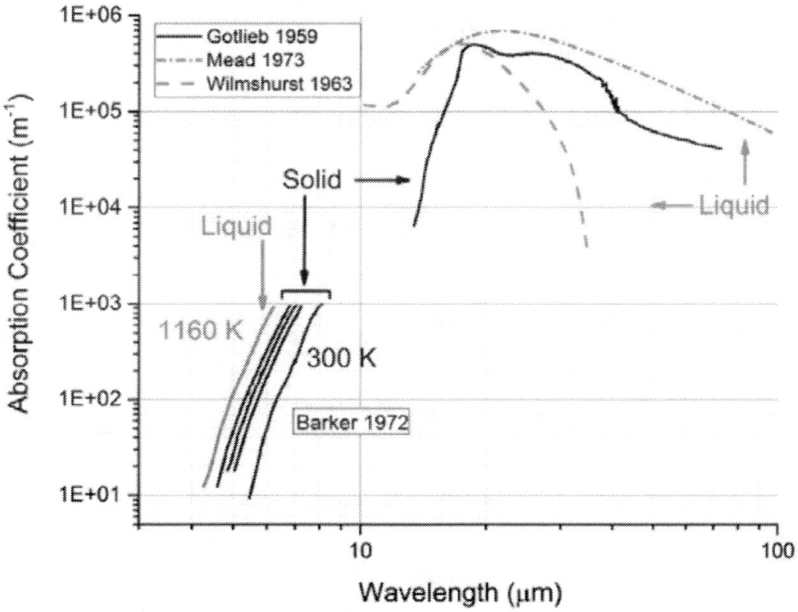

- LiF IR absorption well studied over wide range of wavelengths and temperatures
- Liquid to solid phase change does not produce major changes

[4]D. Mead, 'Comparison of the Optical and Dielectric Properties of Crystalline and Molten Lithium Fluoride', J. Phys. C Solid State Phys., vol. 7, pp. 445-453, (1973);[5]J. K. Wilmshurst, 'Infrared Spectra of Molten Salts', J. Chem. Phys., vol. 39, no. 10, pp. 2545-2548, (1963);[6]A. J. Barker, 'Effect of Melting on Multiphonon Infrared Absorption Spectra of KBr, NaCl and LiF', J. Phys. Part C Solid State Phys., vol. 5, no. 16, p. 2276-, (1972);[7]M. Gottlieb, 'Optical Properties of Lithium Fluoride in the Infrared', J. Opt. Soc. Am., vol. 50, no. 4, p. 343, (1960)

WILL DERDEYN
HEATANDMASS.EP.WISC.EDU 9

Extrapolation of Data Fit

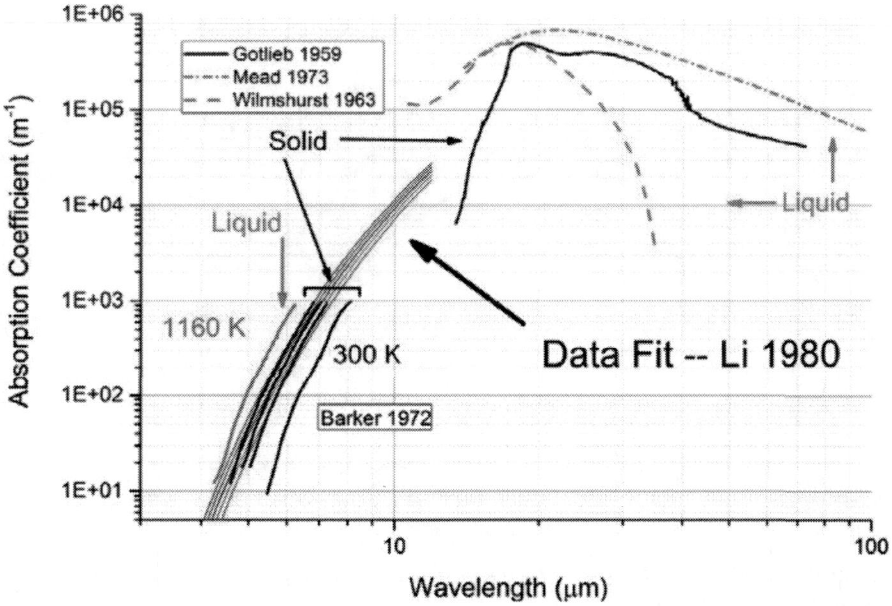

- Li compiled extensive data on alkali halides, observed an exponential behavior at the multiphonon edge, and constructed a data fitting equation [9]
- LiF fit only strictly valid for 4.17 to 8.33 μm range, no data to support extrapolation

[4]D. Mead, 'Comparison of the Optical and Dielectric Properties of Crystalline and Molten Lithium Fluoride', J. Phys. C Solid State Phys., vol. 7, pp. 445-453, (1973);[5]J. K. Wilmshurst, 'Infrared Spectra of Molten Salts', J. Chem. Phys., vol. 39, no. 10, pp. 2545-2548, (1963);[6]A. J. Barker, 'Effect of Melting on Multiphonon Infrared-Absorption Spectra of KBr, NaCl and LiF', J. Phys. Part C Solid State Phys., vol. 5, no. 16, p. 2276-, (1972);[7]M. Gottlieb, 'Optical Properties of Lithium Fluoride in the Infrared', J. Opt. Soc. Am., vol. 50, no. 4, p. 343, (1960);[8]H. H. Li, 'The Infrared Absorption Coefficient of Alkali Halides', Int. J. Thermophys., vol. 1, no. 1, pp. 97-134, (1980)

WILL DERDEYN
HEATANDMASS.EP.WISC.EDU 10

BeF_2 Transmission Spectra

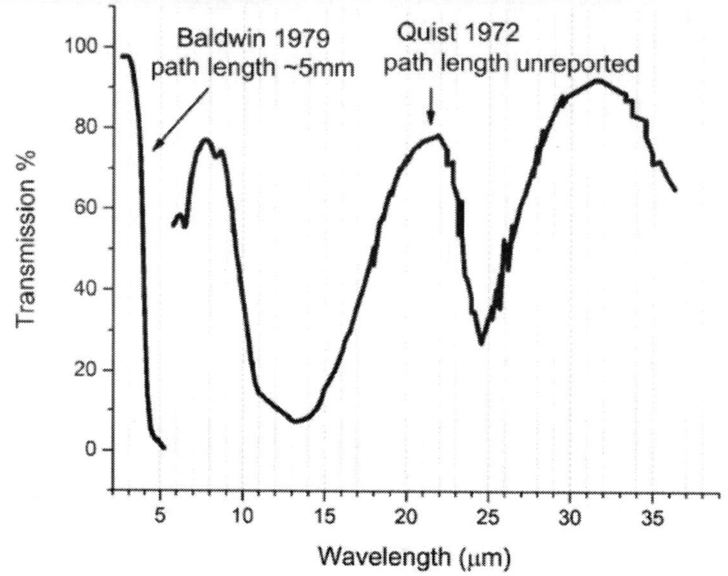

- Baldwin collected the transmission spectrum for the multiphonon edge of BeF_2 glass [10]
- Quist collected transmission spectra of ground BeF_2 glass with same experimental description reported as for FLiBe data [4]

[9]C. M. Baldwin and J. D. Mackenzie, 'PREPARATION AND PROPERTIES OF WATER-FREE VITREOUS BERYLLIUM FLUORIDE', J. Non. Cryst. Solids, **vol. 31**, pp. 441–445, (1979). [10] S. Quist, J. B. Bates, et al., 'Raman spectra of vitreous, polycrystalline, and molten BeF2 to 630°C and the infrared spectrum of vitreous BeF2 at 25°C', Spectrochim. Acta, **vol. 28A**, no. 1973, pp. 1103–1105, (1972).

WILL DERDEYN
HEATANDMASS.EP.WISC.EDU

11

BeF_2 Absorption

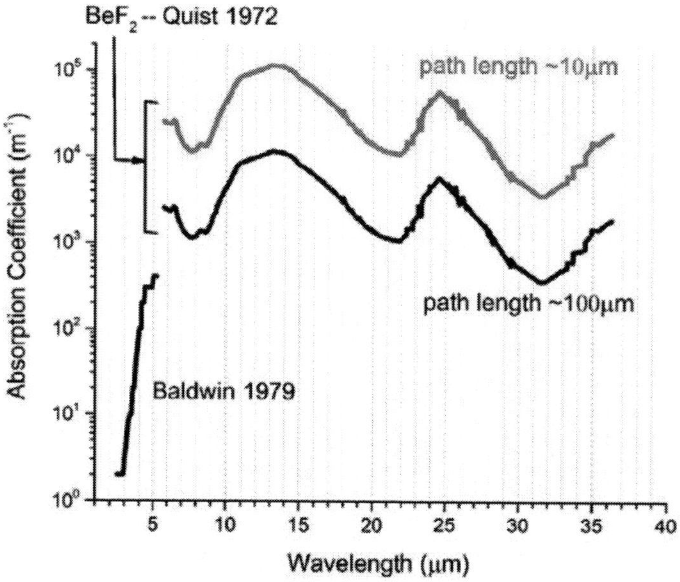

- We calculated absorption coefficients for reported and assumed path lengths of Baldwin's and Quist's data, respectively
- Less data to extrapolate than for LiF, but more uncertainty in peak location

[9]C. M. Baldwin and J. D. Mackenzie, 'PREPARATION AND PROPERTIES OF WATER-FREE VITREOUS BERYLLIUM FLUORIDE', J. Non. Cryst. Solids, **vol. 31**, pp. 441–445, (1979). [10] S. Quist, J. B. Bates, et al., 'Raman spectra of vitreous, polycrystalline, and molten BeF2 to 630°C and the infrared spectrum of vitreous BeF2 at 25°C', Spectrochim. Acta, **vol. 28A**, no. 1973, pp. 1103–1105, (1972).

WILL DERDEYN
HEATANDMASS.EP.WISC.EDU

12

BeF_2: Data Fitting

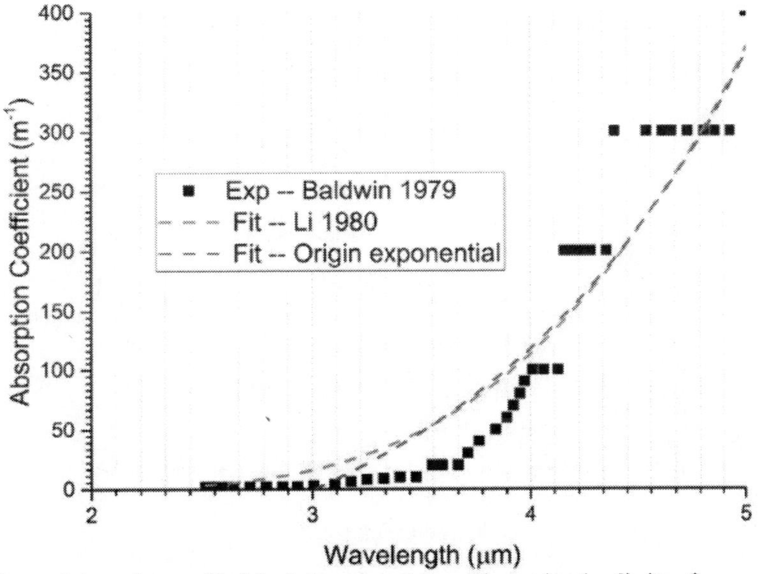

- Lack of precision due to Baldwin's reporting only a single digit of accuracy
- We used Li's fitting equation with alkaline earth parameters to fit Baldwin's data
- We also fit a standard exponential to Baldwin's data

[9]C. M. Baldwin and J. D. Mackenzie, 'PREPARATION AND PROPERTIES OF WATER-FREE VITREOUS BERYLLIUM FLUORIDE', J. Non. Cryst. Solids, **vol. 31**, pp. 441–445, (1979).[11]H. H. Li, 'Absorption Coefficient of Alkaline Earth Halides', West Lafayette, IN, (1980).

Extrapolation of Data Fits

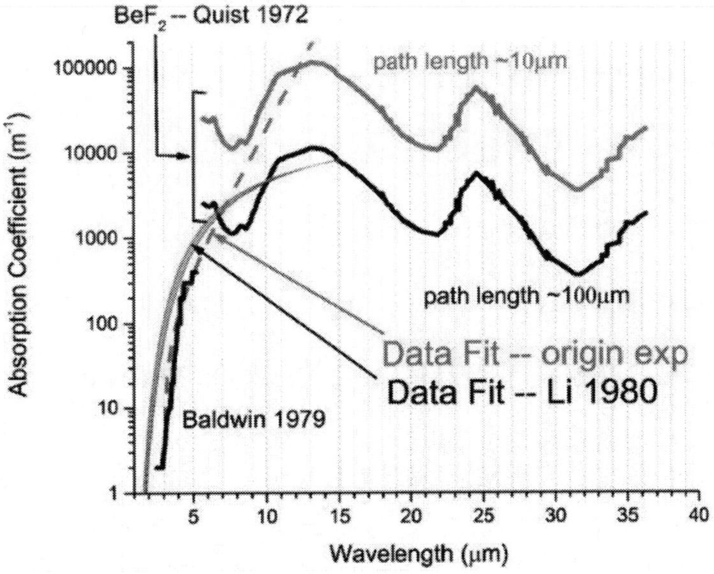

- The two fits agree qualitatively with the upper and lower bounds for BeF_2 absorption

[9]C. M. Baldwin and J. D. Mackenzie, 'PREPARATION AND PROPERTIES OF WATER-FREE VITREOUS BERYLLIUM FLUORIDE', J. Non. Cryst. Solids, **vol. 31**, pp. 441–445, (1979).[11]H. H. Li, 'Absorption Coefficient of Alkaline Earth Halides', West Lafayette, IN, (1980).
[10] S. Quist, J. B. Bates, et al., 'Raman spectra of vitreous, polycrystalline, and molten BeF2 to 630°C and the infrared spectrum of vitreous BeF2 at 25°C', Spectrochim. Acta, **vol. 28A**, no. 1973, pp. 1103–1105, (1972).

Estimated FLiBe Spectra

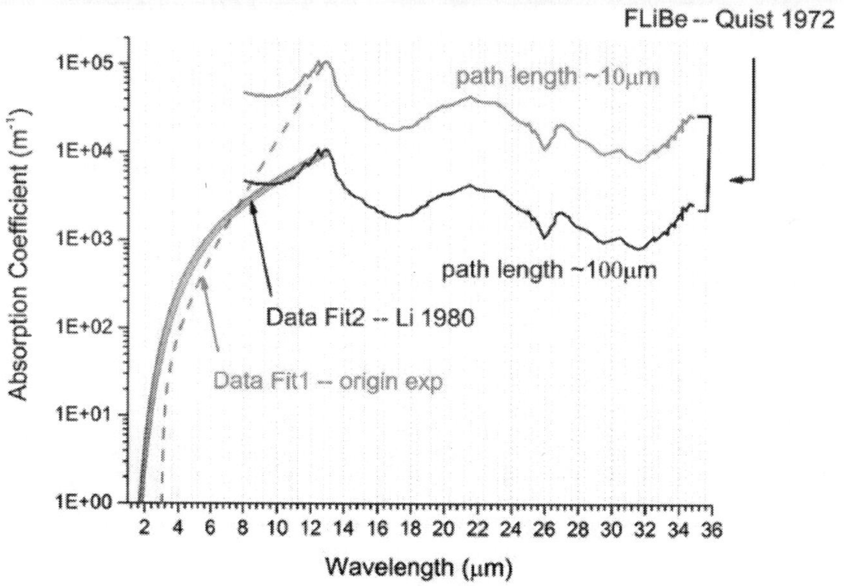

- For both cases, BeF_2 absorption spectra was averaged with LiF using the formula of Chaleff[2]
- Qualitative agreement present still between calculated edge and expected peaks

[2]E. S. Chaleff, 'The Radiative Heat Transfer Properties of Molten Salts and Their Relevance to the Design of Advanced Reactors', (2016). [1]H. H. Li, 'Absorption Coefficient of Alkaline Earth Halides', West Lafayette, IN, (1980). [3] S. Quist, J. B. Bates, et al., 'Raman spectra of vitreous, polycrystalline, and molten BeF2 to 630°C and the infrared spectrum of vitreous BeF2 at 25°C', Spectrochim. Acta, vol. 28A, no. 1973, pp. 1103–1105, (1972).

WILL DERDEYN
HEATANDMASS.EP.WISC.EDU
15

Outline

1. Motivation
2. Background Physics
3. Review of FLiBe Literature Data
4. RHT Analysis
5. Uncertainty
6. Conclusions and Path Forward

WILL DERDEYN
HEATANDMASS.EP.WISC.EDU
16

Absorption Data Truncation

- Optical thickness defined as the following:
- $\tau = \int_0^L k ds$

Where:

- ds = the infinitesimal path length
- k = absorption coefficient [=] m^{-1}
- L = relevant length scale [=] m

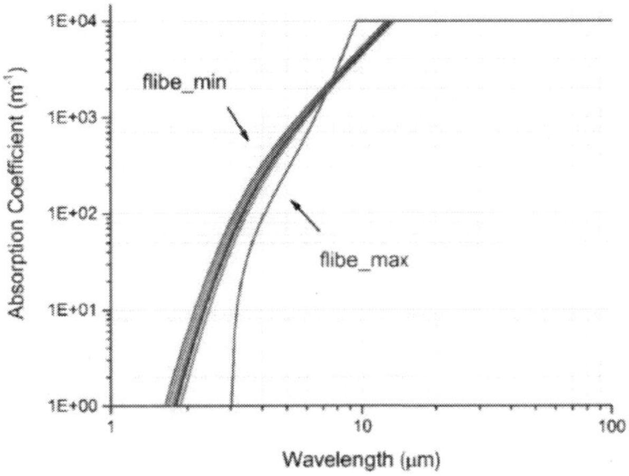

- Assuming k does not vary in space:
- $\tau = \int_0^L k ds = \mathrm{kL}$

- For our analysis, L = 1cm, which is the tube diameter in the TCHX
- Next, we will observe that RHT behavior is insensitive to $\tau > 100$[2].

$$\therefore k_{max} = \frac{\tau_{max}}{L} = \frac{100}{.01\ m} = 1E4\ m^{-1}$$

[12]M. F. Modest, Radiative Heat Transfer. (2003).[2]E. S. Chaleff, 'The Radiative Heat Transfer Properties of Molten Salts and Their Relevance to the Design of Advanced Reactors', (2016).

Optical Thickness Calculation

- We can't immediately plug in our absorption coefficient due to its spectral dependence
- So, for the temperature range of 400 – 800°C, we performed the following integration:

$$\bar{k}(T) = \frac{\int_0^\infty k_\lambda * E_{bb} * d\lambda}{\int_0^\infty E_{bb} * d\lambda} = \frac{\int_0^\infty k_\lambda * E_{bb} * d\lambda}{\sigma T^4} \cong \frac{\int_0^{\lambda max} k_\lambda * E_{bb} * d\lambda}{\sigma T^4}$$

Assumption that k goes to zero beyond λ_{max} = 100 μm

$$\text{where } E_{bb} = \frac{2*h*c^2}{\lambda^5} * \frac{1}{e^{\frac{hc}{\lambda k_B T}} - 1}$$

Stefan–Boltzmann Law

h = 6.626e-34 m^2kg/s
c = 3e8 m/s
k = 1.38e-23 m^2kg/s^2K
σ = 5.67e-8 W/m^2*K^4

- With this spectrally averaged value of $\bar{k}(T)$ in hand, we then calculated optical thickness as follows:
- $\tau = \int_0^L \bar{k} ds = L\bar{k} = (.01\mathrm{m}) * \bar{k}$

[12]M. F. Modest, Radiative Heat Transfer. (2003).

FLiBe Optical Thickness for FHR

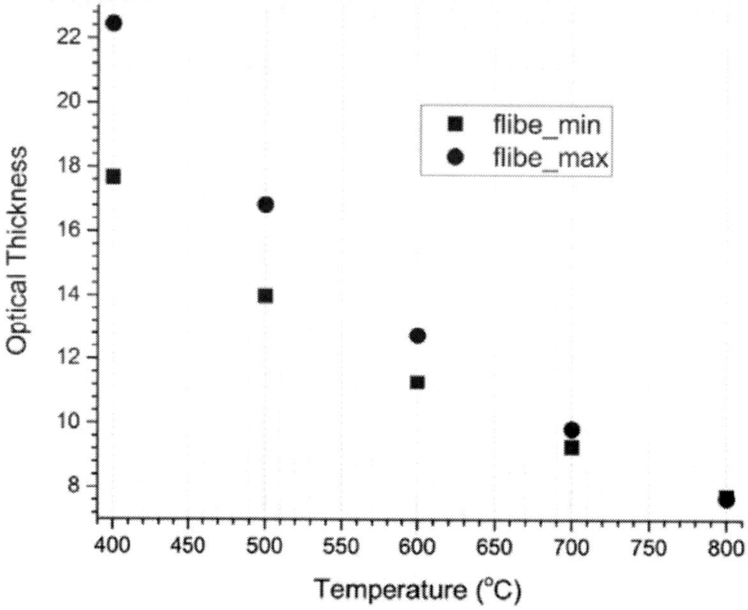

- Results show that FLiBe will behave increasingly transparent as temperature increases
- Uncertainty due to far infrared data has less of an effect as well

WILL DERDEYN
HEATANDMASS.EP.WISC.EDU
19

Outline

1. **Motivation**
2. **Background Physics**
3. **Review of FLiBe Literature Data**
4. **RHT Analysis**
5. **Uncertainty**
6. **Conclusions and Path Forward**

WILL DERDEYN
HEATANDMASS.EP.WISC.EDU
20

Effect of Mixing: Quasi-lattice

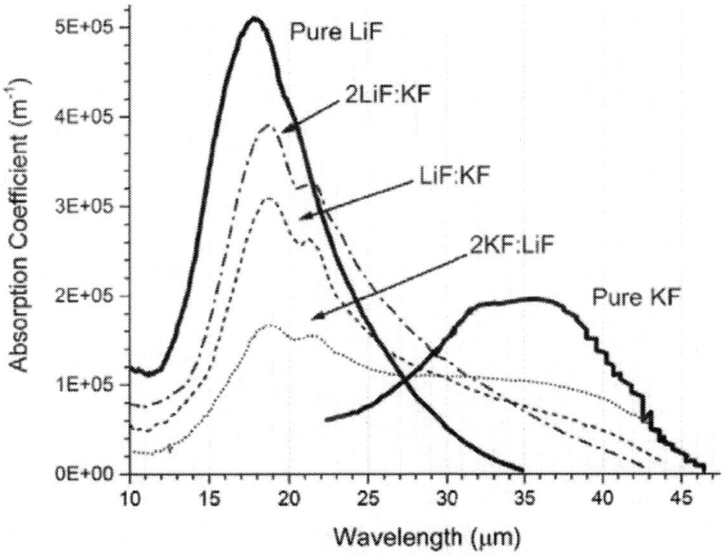

- Wilmshurst collected data on pure LiF, KF and their mixtures[6]
- He suggested peaks were result of 'quasi-lattice' vibrations within the melt
- He observed non-linear behavior during transition between pure compounds

[5]J. K. Wilmshurst, 'Infrared Spectra of Molten Salts', J. Chem. Phys., vol. 39, no. 10, pp. 2545–2548, (1963).

WILL DERDEYN
HEATANDMASS.EP.WISC.EDU
21

Effect of Mixing: Complex Ion

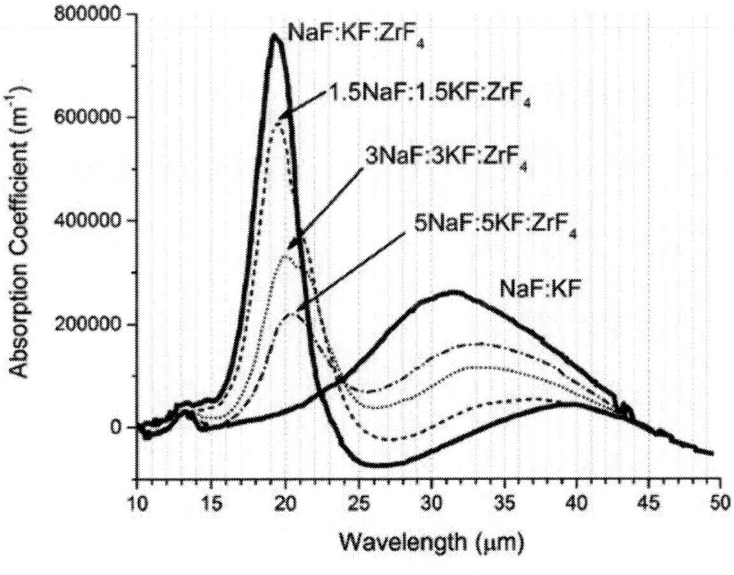

- Wilmshurst also collected data on NaF, KF, and ZrF_4 mixtures[6]
- He suggested a transition from ZrF_4 rich glassy lattice, to a 'hard-sphere' quasi lattice of NaF:KF

[5]J. K. Wilmshurst, 'Infrared Spectra of Molten Salts', J. Chem. Phys., vol. 39, no. 10, pp. 2545–2548, (1963).

WILL DERDEYN
HEATANDMASS.EP.WISC.EDU
22

ZrF_6 Complex Ion

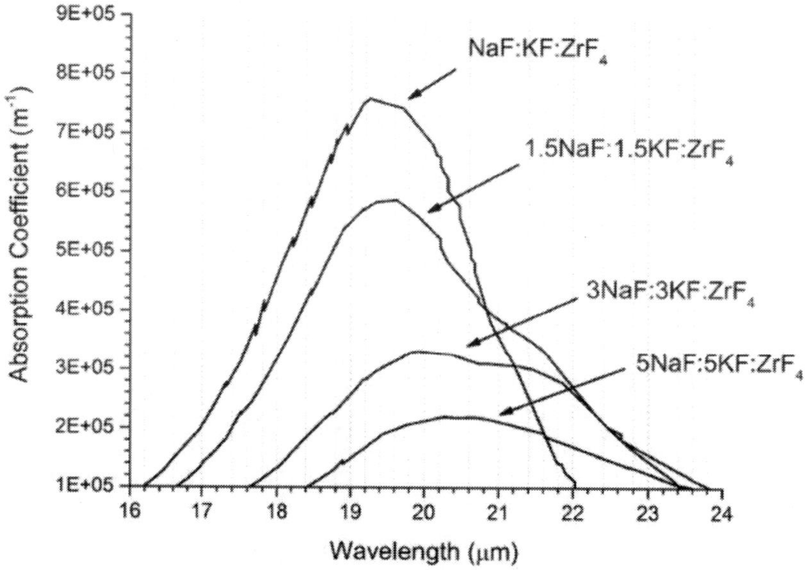

- Zooming in on the ZrF_4 peak reveals an interesting transition
- Wilmshurst suggested the formation of ZrF_6 as an intermediate species between the glassy ZrF_4 rich and quasi-lattice ZrF_4 poor mixtures

[5] J. K. Wilmshurst, 'Infrared Spectra of Molten Salts', J. Chem. Phys., vol. 39, no. 10, pp. 2545–2548, (1963).

Outline

1. **Motivation**
2. **Background Physics**
3. **Review of FLiBe Literature Data**
4. **RHT Analysis**
5. **Uncertainty**
6. **Conclusions and Path Forward**

Conclusions

- IR absorption literature data has been collected for:
  - LiF
  - BeF_2
  - FLiBe
  - LiF:KF mixtures
  - NaF:KF:ZrF_4 mixtures
- An estimate for the IR absorption coefficient of FLiBe has been generated
- The optical thickness of FLiBe in TCHX at 800°C has been found to converge to 8

Next Steps

- Continue work on experimental setup
- Validate RHT analysis

497245 Effect of LiOH Contamination on the Electrolytic Reduction of Spent Oxide Fuel in Molten LiCl-Li_2O

Abstract:

Spent oxide fuel must be reduced to metallic form prior to electrorefining. In the oxide reduction process, spent UO_2 based fuel is contained in a stainless steel cathode basket and immersed into an electrochemical cell containing a LiCl-Li_2O melt at 650°C, at which point oxide reduction can be achieved via direct electrolytic reduction (DER) of oxides with simultaneous evolution of O_2 gas at the anode. Both Li_2O and UO_2 can be reduced in the cathode basket. Lithium metal formed from reduction of Li_2O will spontaneously reduce UO_2, so the net result is the same. However, moisture contamination in the salt results in the formation of LiOH, which interferes with the reduction of UO_2 and Li_2O and decreases current efficiency. LiOH reduction occurs at a less cathodic potential than does Li_2O or UO_2. In this study, the impact of moisture and LiOH on the oxide reduction process and the effect of contamination removal methods were investigated. Thermogravimetric analysis (TGA) was used to measure moisture content in salts. Reduction peaks in cyclic voltammetry (CV) scans were associated with the presence of LiOH in the salt. It was found that LiOH could be removed from the salt via addition of Li metal or electrochemically through chronopotentiometric methods. Development and optimization of a salt cleanup process will be discussed as an important requirement to efficiently running DER of spent oxide fuel.

497429 Effect of Rotation on Li_2O Entrainment in a Uranium Oxide Reduction Process

Abstract:

To apply pyroprocessing to oxide used fuel, an oxide reduction step must precede electrorefining. The current leading technology for oxide reduction is a high temperature electrolytic process carried out in molten lithium chloride salt. In this process, the used fuel is loaded into cathode baskets. Lithium oxide entrainment in these cathode baskets is problematic, as it can detrimentally affect the electrorefining process if it carries over with the reduced fuel. Gaining an understanding of the operating parameters that effect lithium oxide entrainment in the cathode basket is the goal of the research reported in this paper. In particular, the effect of basket rotation on lithium oxide entrainment has been studied. A special cathode basket has been designed, fabricated, and used in small-scale electrolytic reduction tests of U_3O_8 and UO_2. The basket directly polarizes the uranium oxide particles, while its external walls are electrically isolated. Results of constant current reduction tests at various rates of rotation will be presented. A method for distinguishing between lithium oxide in different environments (inside and outside of the uranium particles) will be presented.

488466 Thermal Conductivity of Tungsten: Effects of Plasma-Related Structural Defects from Molecular-Dynamics Simulation

Abstract:

The high thermal conductivity of tungsten (175 W/m·K at room temperature) is one of the main reasons for its consideration as a plasma-facing material (PFM) in nuclear fusion reactors. However, the exposure of PFMs to large fluxes of low-energy helium (He) irradiation from the plasma causes significant microstructural changes in the PFM's near-surface region. Structural defects such as He nanobubbles due to plasma exposure are expected to have significant impact on tungsten's thermal conductivity in the near-surface region, as a result of the stronger electron and phonon scattering mechanisms by the plasma-related structural defects.

In this presentation, we report on the effects of plasma-related structural defects on the lattice, i.e., phonon, thermal conductivity of tungsten based on an atomic-scale computational analysis. We have conducted a systematic study of thermal transport using non-equilibrium molecular-dynamics (NEMD) simulations and computed the lattice thermal conductivities of tungsten single crystals and tungsten single crystals containing nanoscale-size pores or voids and He nanobubbles as a function of void/bubble size and gas pressure in the He bubbles. The calculated lattice thermal conductivities of tungsten single crystals range from 15.0 to 15.6 W/m·K along different heat flux directions, which is, as expected for a metal, just under 10% of the overall thermal conductivity of tungsten. Heat fluxes along the [100], [110], [111], and [211] crystallographic directions have been examined, and the weak dependence of the lattice thermal conductivity on the heat flux direction implies a weak anisotropy in heat conduction in tungsten. The presence of nanoscale voids in the crystal causes a significant reduction in its lattice thermal conductivity, which decreases with increasing void size. Filling the voids with He to form He nanobubbles and increasing the bubble pressure leads to further reduction, up to ~80%, of the tungsten's lattice thermal conductivity, while the anisotropy in heat conduction remains weak for tungsten single crystals containing nanoscale size voids and He nanobubbles throughout the pressure range examined.

In addition, we have calculated the pressure distribution in the crystalline region that surrounds the He nanobubbles and analyzed the phonon transport and scattering mechanisms that mediate heat conduction in tungsten. We have found that the significant reduction of tungsten's lattice thermal conductivity in the region that contains He nanobubbles is due to phonon scattering from the nanobubbles, as well as the deformation of the lattice around the nanobubbles and the formation of lattice imperfections in these regions at higher bubble pressure.

511784 The Heat and Mass Transport Questions in Molten Fluoride Salts that are Relevant to Advanced Nuclear Reactors

Abstract:

Molten salt technology that has the possibility of being a sustainable source of energy that does not produce atmospheric pollution and that is competitive with energy from fossil fuels. There are more than a dozen start-up companies pursuing molten salt technology, both U.S. and internationally. Some are driven by sustainability characteristics of molten salt reactors, which can recycle spent fuel from conventional nuclear reactors. Other companies are driven by the economic advantages that arise from the inherent safety features of molten salts as coolants for nuclear reactors. For example, molten salt mixtures have low vapor pressure and can operate at atmospheric pressure without boiling up to very high temperature (1000 °C and above). The atmospheric pressure operation significantly reduces costs and material testing timelines for advanced nuclear reactors. The high temperature operation enables coupling to high-efficiency modern gas turbines (open air Brayton cycles with combined cycle efficiencies of >60%), compared to steam turbines that are used by conventional, water-cooled nuclear power plants that operate below 350 °C (with efficiencies of ~35%). The high temperature operation also enables a broader market for process heat applications, and the use of gas turbines enables the production of rapidly-deployable peaking power by the use of gas co-firing.

Some of the technical challenges important in the commercialization of advanced nuclear reactors that use molten salts are tritium management and more generally containment of radionuclides, control of actinide inventory in the liquid phase, and management of transients that involve liquid-solid phase changes. This talk will provide an overview of ongoing research at the University of Wisconsin on heat and mass transport phenomena in molten fluoride salts.

492574 Effect of Process Parameters on Chlorination of Metallic Rare Earth/Actinide Mixtures

Abstract:

Separation of actinides from rare earth metals has benefits in metallurgy for rare earth extraction as well as nuclear for reprocessing nuclear materials. Efficient separation is challenging, because of the similarities between actinides and lanthanides. Traditional aqueous solvent extraction methods are often undesirable because of the need to dispose of large volumes of toxic, radioactive solvents in addition to the threat for their application to proliferation. A two-step process is being developed to prepare actinide/rare earth solid metal samples for purificationâ€"featuring hydriding followed by chlorination. Results for experiments are reported in which metal samples containing cerium, uranium, and transition metals were hydrided with H_2 followed by chlorination with anhydrous Cl_2. Results will be reported for varying parameters such as temperature, exposure time, and gas dilution to determine the effect of these parameters on metal conversion to chlorides. Chloride ion selective electrodes are the primary method for measuring the extent of chlorination of the metal samples.

491651 Tritium Management Approaches and Their Impacts on Tritium Distribution within a Unf Reprocessing Plant

Abstract:

During the reprocessing of used nuclear fuel (UNF), volatile radionuclides are released from the fuel matrix and will be discharged from the facility stack if no recovery processes are in place to limit their release. The volatile radionuclides of concern are ^{3}H, ^{14}C, ^{85}Kr, and ^{129}I.

While iodine is typically the focus of most off-gas capture studies due to the high decontamination factor (DF) requirements and multiple release pathways, prior studies have indicated that the control of tritium could also be a significant issue for future reprocessing plants, especially if relatively short-cooled fuels are being processed [1,2]. A previous study by Jubin and Spencer [3] examined the accumulation and distribution of tritium in an aqueous-based reprocessing plant as well as the impacts of various approaches to remove the tritium from the aqueous and off-gas streams. The evaluated off-gas streams included the dissolver off-gas (DOG), the vessel off-gas (VOG), and the melter off-gas (MOG). The extent to which the various streams must be treated to achieve the overall specified facility tritium DFs was examined. The previous study assumed that the head-end of the UNF reprocessing plant was a typical chop/leach operation in which no effort is made to remove the tritium from the fuel prior to dissolution.

The current study expands upon that work with the addition of a tritium pretreatment operation in which most, if not all, of the tritium is volatilized from the fuel prior to the dissolution step. Thus the amount of tritium that reaches the aqueous portion of the plant is minimized. This study examines the impact of incomplete volatilization of tritium from the fuel, the impact of tritium pretreatment on tritium concentrations within the plant, and the estimated volumes of tritiated waste that would be generated. Comparisons are made with the waste generated in two limiting cases: complete volatilization of the tritium in the fuel and no volatilization of the tritium. In all cases, comparable levels of tritium controls are utilized to limit the emissions to provide the same overall plant-wide tritium DF.

1. R. JUBIN, N. SOELBERG, D. STRACHAN, and G. ILAS, *Assessments and Options for Removal and Immobilization of Volatile Radionuclides from the Processing of Used Nuclear Fuel*, FCR&D-SWF-2011-000305, UT-Battelle, LLC, Oak Ridge National Laboratory (2011).

2. R. JUBIN, N. SOELBERG, D. STRACHAN, and G. ILAS, *Fuel Age Impacts on Gaseous Fission Product Capture During Separations*, FCRD-SWF-2012-000089, PNNL-22550, UT-Battelle, LLC, Oak Ridge National Laboratory (2012).

3. R. T. JUBIN and B. B. SPENCER, "Tritium Distribution and Capture Requirements for UNF Reprocessing," *Transactions of the American Nuclear Society*, Vol. 114, New Orleans, La. (June 12–16, 2016).

This manuscript has been authored by UT-Battelle, LLC, under Contract No. DE-AC0500OR22725 with the U.S. Department of Energy. The United States Government retains and the publisher, by accepting the article for publication, acknowledges that the United States Government retains a nonexclusive, paid-up, irrevocable, worldwide license to publish or reproduce the published form of this manuscript, or allow others to do so, for the United States Government purposes. The Department of Energy will provide public access to these results of federally sponsored research in accordance with the DOE Public Access Plan (http://energy.gov/downloads/doe-public-access-plan).

496613 Effective Removal of Pertechnetate (TcO_4^-), Iodide (I^-) and Iodate (IO_3^-) from Groundwater By Organoclays and Granular Activated Carbon

Abstract:

^{99}Tc and ^{129}I are major long-life fission products from nuclear power generation. Over the years, Tc and I have been inadvertently introduced into the environment from leaks at waste storage facilities and currently are key risk drivers at the US DOE sites. The most common chemical form of Tc and I in liquid nuclear wastes and in the environment is anionic pertechnetate (TcO_4^-) or iodide (I^-), iodate (IO_3^-) and organo-I. They display limited adsorption onto common sediment minerals and are highly mobile making them difficult to capture or to be immobilized. As the stockpile of ^{99}Tc- and ^{129}I-bearing nuclear waste continues to increase rapidly, novel sequestration technologies are needed to reduce its potential contamination of the environmental and living organisms.

In this work, inexpensive organoclays and granular activated carbon (GAC) were evaluated for remediation of TcO_4^-, I^- and IO_3^- from groundwater using batch experimental set-up under atmospheric conditions. For TcO_4^-, the adsorption coefficient (K_d) values for the organoclays and GAC samples were greater than 1×10^5 mL/g, with nearly 100% adsorption. Their adsorption kinetics was fast, completing within 1 hour. The adsorption capacities were >3.2 mg/g over a wide groundwater pH range and under oxic conditions. For I^- and IO_3^-, the adsorption capacities onto the organoclays and GAC in groundwater under oxic condition were high as well: 30 mg/g for I^-, and >100 mg/g for IO_3^-.

Further, Tc K-edge XANES and EXAFS spectra of the organoclays and GAC after Tc adsorption were collected using the beamline 20-BM at the Advanced Photon Source (APS) in order to understand Tc speciation and binding chemistry. Both Tc K-edge XANES and EXAFS spectra indicated that the Tc species bound onto these sorbents is TcO_4^-, rather than its reduced Tc(IV). TcO_4^- is likely bonding with N ligands present in functionalized organic molecules of organoclays or other surface moieties present on GAC. The innovation of this technology is that inexpensive organoclays and GAC can sequester Tc contaminant in the form that it exists naturally in the environment, and an artificial reducing environment is not needed to be created. These results provide an applied solution for treating TcO_4^-, I^- and IO_3^-contamination in the environment or in nuclear wastes.

Crossflow Filtration Fouling Behavior at Low Solids Concentration

PP Schonewill

RC Daniel
CA Burns
SD Hoyle
Pacific Northwest National Laboratory
31 October 2016

2017 AIChE Annual Meeting, Session 327 PNNL-SA-130150

Motivation

- Crossflow filtration: mature and robust waste separation technology
  - Deployed at Savannah River Site, West Valley
  - Proposed for use at Hanford waste processing facilities
- Near-term proposed waste solids removal processes expected to be conducted at low solids concentration

Interior of filter surface

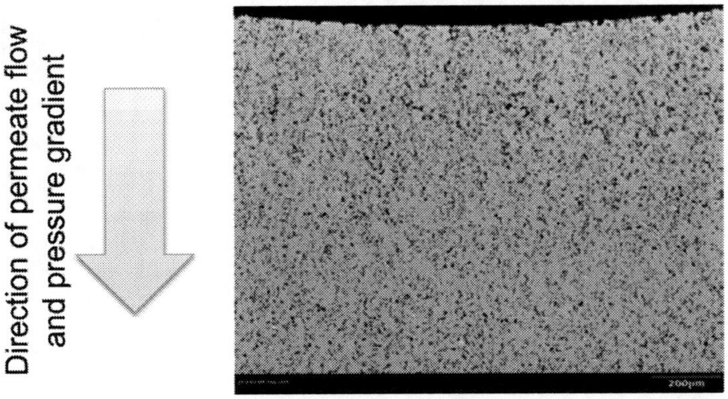

Potential Application with Low UDS

Pacific Northwest NATIONAL LABORATORY

- Low-Activity Waste Pretreatment System (LAWPS): UDS are anticipated to be low
  - Feed is decanted from DSTs that have been allowed to settle
  - UDS specified as < 3 wt%, and in most cases, << 1 wt%.
- Historical crossflow filter test data with Hanford wastes or simulants focused on high level waste with UDS > 3 wt%
- Uncertainty in flux performance and the effectiveness of backpulsing and cleaning using process fluids

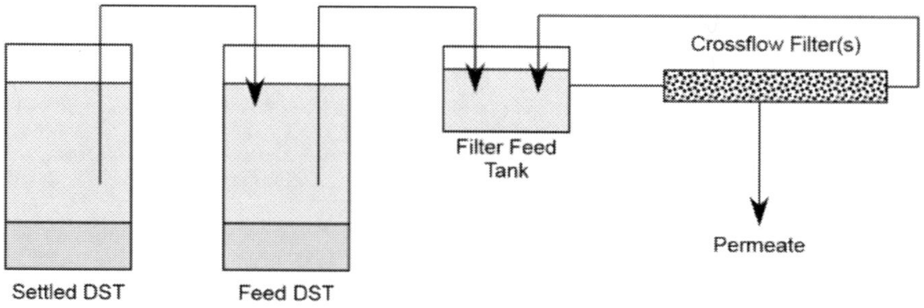

3

31 October 2017

Crossflow Filter Media

Pacific Northwest NATIONAL LABORATORY
Proudly Operated by Battelle Since 1965

- Crossflow filter (CFF) element
  - Sintered stainless steel
  - Mott Corporation, 0.1 grade
  - Average pore opening 1 to 3 μm (manufacturer data)
  - Rejects solids by tortuosity
  - SEM images (at right) indicate some depressions/openings > 3 μm
- Formation of protective cake unlikely at low solids concentrations
- Higher risk of depth fouling

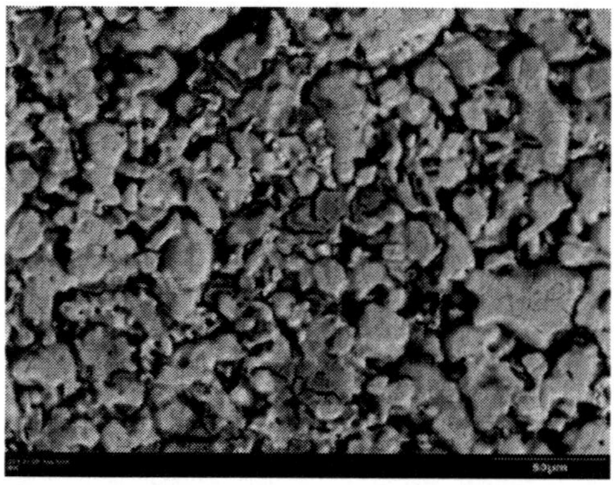

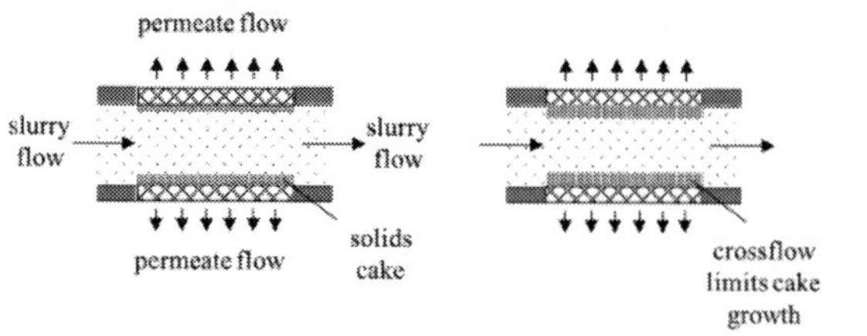

4

31 October 2017

Crossflow Filtration (CFF) Skid

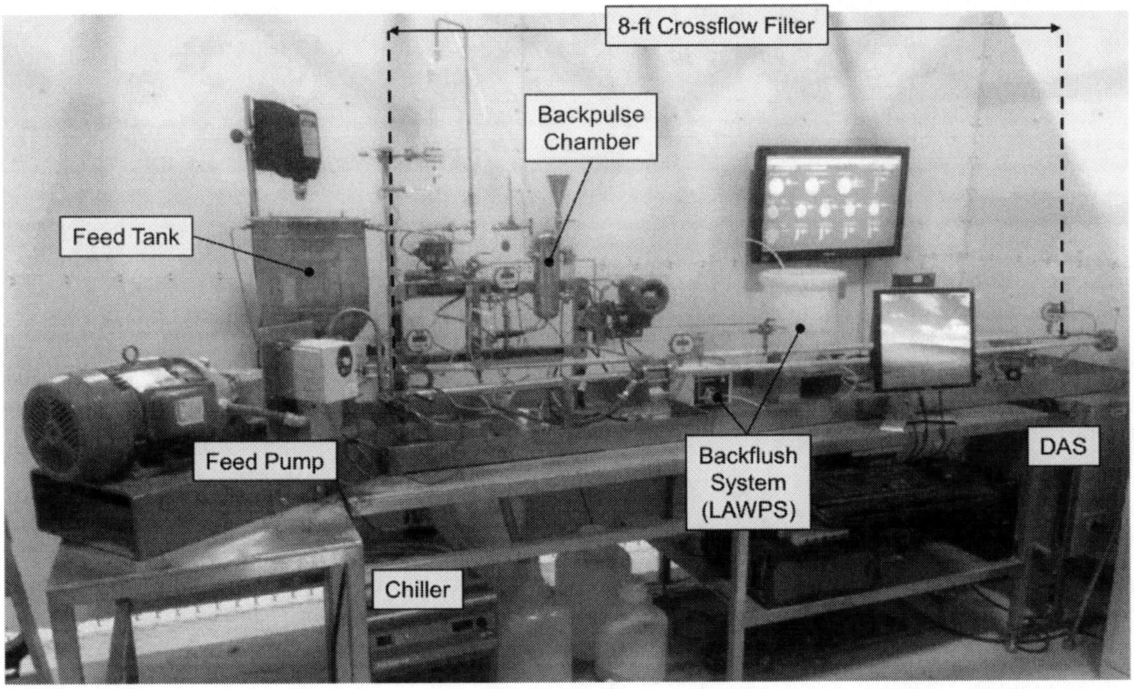

CFF Simulants

- Single solids components in 5.6M sodium supernate
  - Density: 1.26 g/mL
  - Viscosity: 3.4 cP at 25 degrees Celsius

Test	Solid Component	Solids Loading	Temperature
1	Sodium Oxalate	0.001 wt%	20 °C
2	Sodium Phosphate	0.1 wt%	45 °C
3	Iron Oxyhydroxide	0.1 wt%	20 °C
4	Iron Phosphate	0.1 wt%	20 °C
5	Cancrinite	0.1 wt%	20 °C
6	HEDTA	0.1 wt%	20 °C
7	Degraded sRF	~0.1 wt%	20 °C
8	Sodium Fluoride	6 g/L	45 °C

CFF Test Evolution

Pacific Northwest NATIONAL LABORATORY

- Segments of Operation:
  - A – 120 hr baseline period at constant TMP
  - B – 24 hr period after single backpulse
  - C – 8 hr period of backpulsing every 30 min, followed by 16 hr period of undisturbed operation
  - D – In-situ cleaning process using LAWPS planned protocol
  - E – Post-cleaning 24 hr filtration period
  - F – Post-test cleaning and clean water flux measurements

Iron Phosphate Test

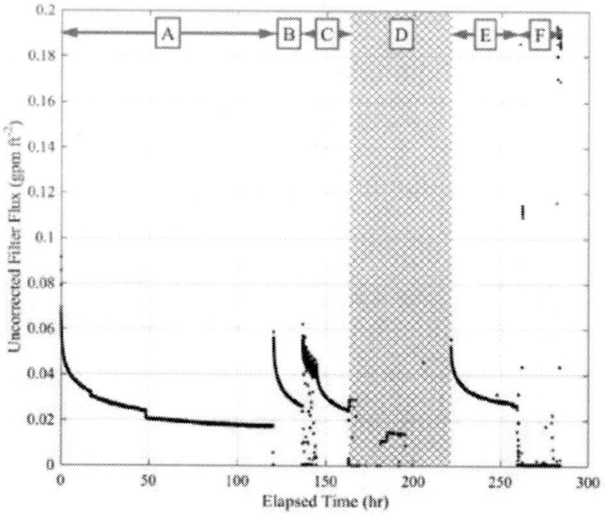

7

31 October 2017

CFF Test Data: 120 Hour Period

Pacific Northwest NATIONAL LABORATORY

Proudly Operated by Battelle Since 1965

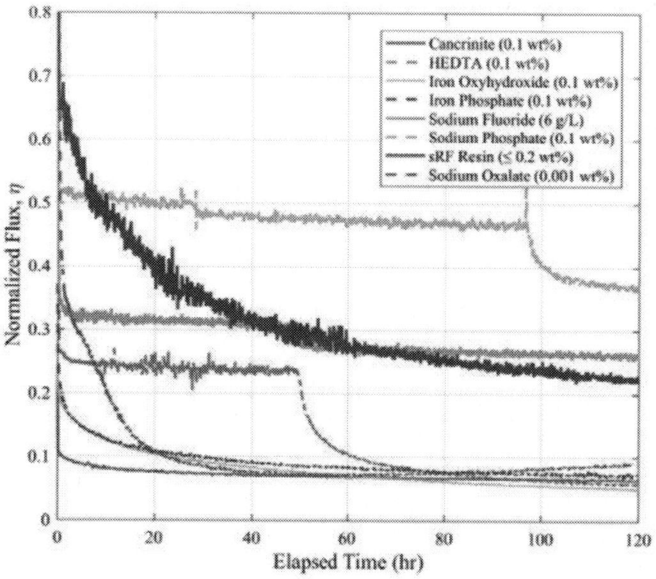

Flux is normalized by the flux expected through a clean filter at the filtration conditions (pressure, viscosity).

δP and μ are both function of time (μ changes with temperature)

$$\eta = \frac{j}{j_o} = j\left(\frac{\mu \Delta_m r_{m,o}}{\delta P}\right)$$

8

31 October 2017

CFF Test Data: "Worst" Foulants

Pacific Northwest
NATIONAL LABORATORY
Proudly Operated by Battelle Since 1965

Initial hour vs. final hour of η, four "worst" foulants

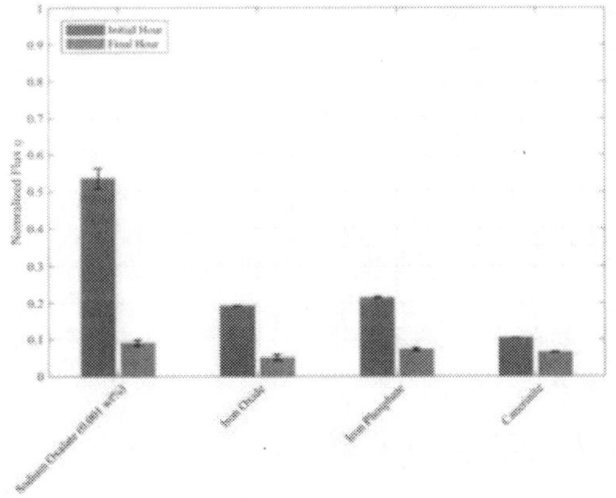

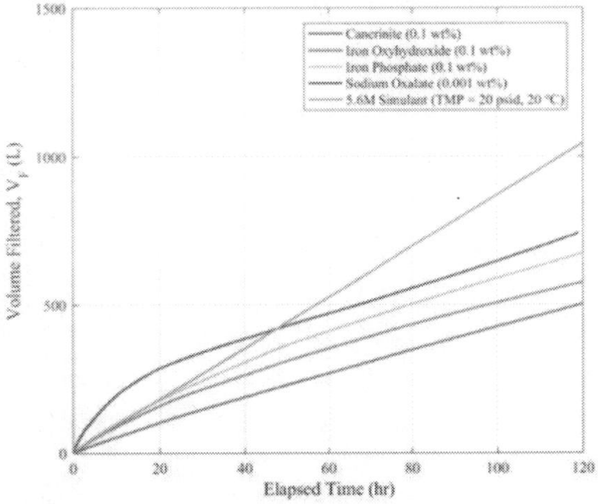

5.6M simulant had solids loading of ~0.45 wt%, all four of the worst foulants significantly degraded performance by comparison.

CFF Test Data: Recovery Operations

Pacific Northwest
NATIONAL LABORATORY
Proudly Operated by Battelle Since 1965

Average normalized flux, initial hour

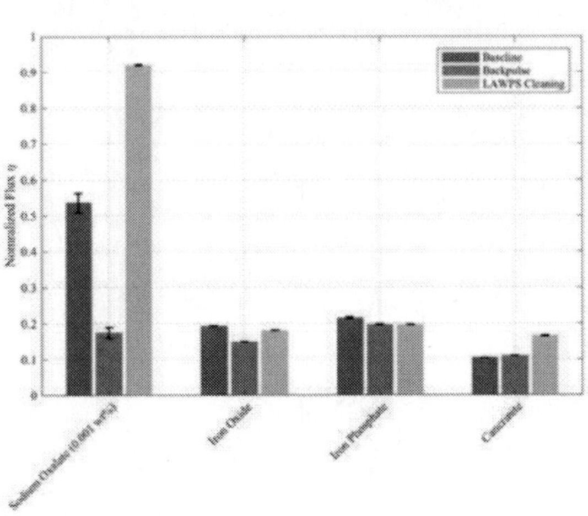

16 hour overlays, iron oxyhydroxide test

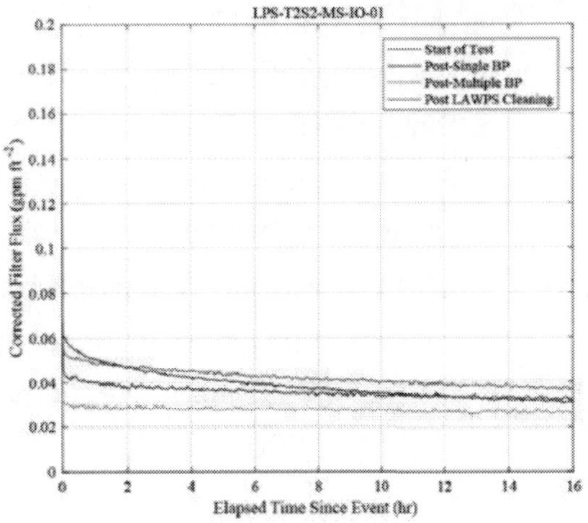

CFF Test Data Comparison

Pacific Northwest NATIONAL LABORATORY

Proudly Operated by Battelle Since 1965

Comparison of simulants with similar solids but a factor of ~20 greater UDS concentration. The data is adjusted to compare on the same basis.

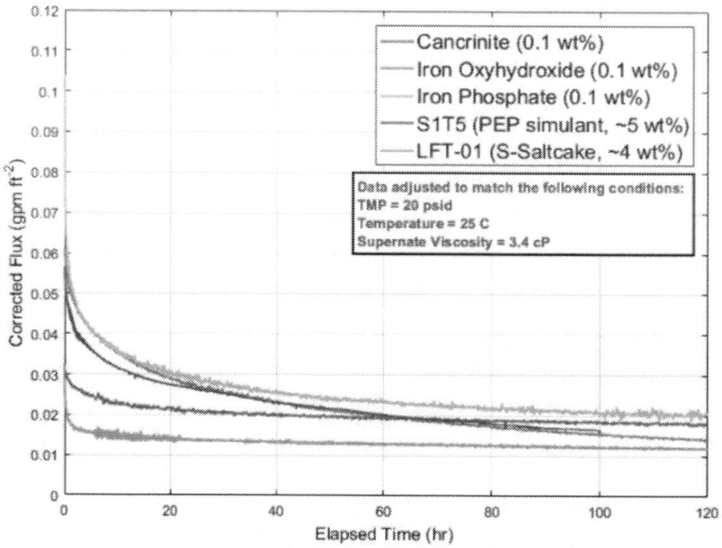

11

31 October 2017

Filter-Particle Surface Interactions

Pacific Northwest NATIONAL LABORATORY

Proudly Operated by Battelle Since 1965

- Filter fouling is significant, even at very low solids concentrations used in this study
- How do particle-filter surface interactions control performance for the system of interest? Namely:
  - Hanford waste (or simulant) chemistry
  - Sintered metal filter
  - Fully turbulent flow
  - Dilute suspensions
- Filtration operation is a black box
  - Can we improve on empirical approach?
  - Knowledge of filter interactions guide operational approaches

12

31 October 2017

Crossflow Visualization Cell

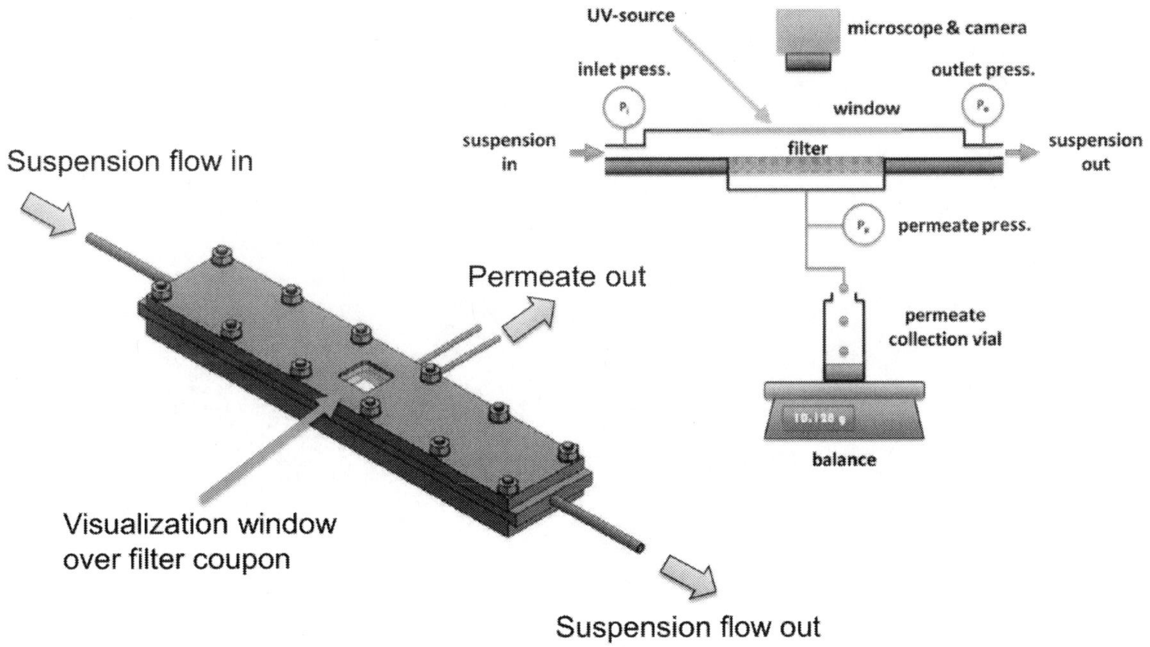

Proof of Concept Testing

Pacific Northwest
NATIONAL LABORATORY
Proudly Operated by Battelle Since 1965

- Aqueous solutions with well-defined solid component:
  - Spherical polyethylene beads
  - Fluoresce under UV light
  - Nominally 30 μm (red) and 5 μm (blue) in size
  - 1 ppmv concentration
- Filter at constant transmembrane pressure
- Collect images of filter surface with time

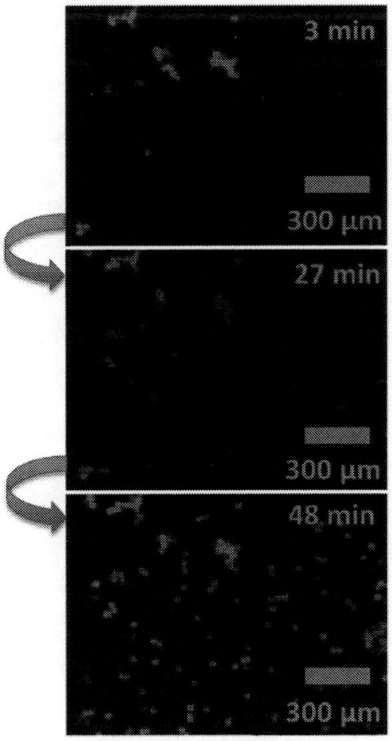

Proof of Concept Data

Normalized flux versus normalized surface coverage
1 ppmv 30 μm beads in water

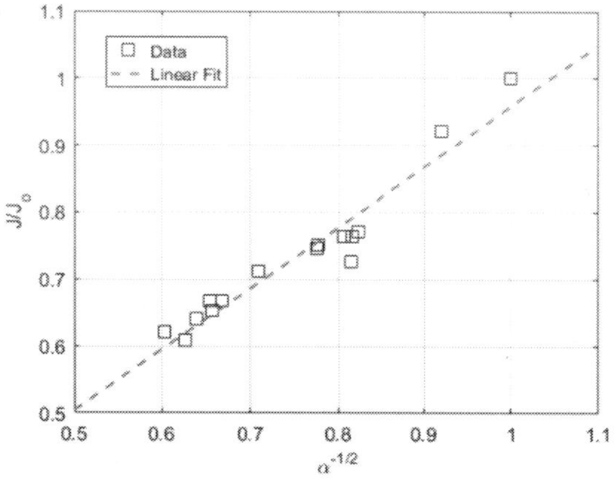

Flux vs. estimated surface coverage
1 ppmv 5 μm beads in water

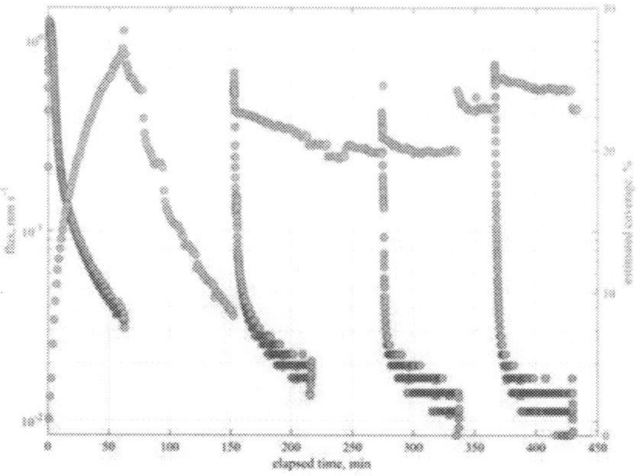

Follow-On Work

- Going beyond proof-of-concept:
  - Bi-disperse bead suspensions
  - "Real" suspensions, i.e. functionalized salts or solid components
  - Effect of concentration
  - Backflushing
  - Laminar versus turbulent flow

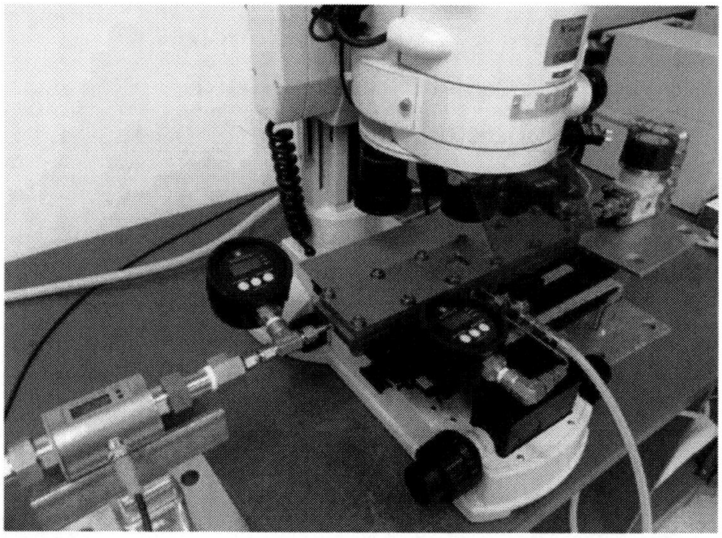

Summary

Pacific Northwest NATIONAL LABORATORY
Proudly Operated by Battelle Since 1965

- Caustic simulants with single solid components at ~0.1 wt%:
  - Generally exhibited fouling that degraded performance
  - Hampered production by a factor of ~2 over an operating period of 5 days for the most severe foulants
  - Exhibited short-term flux recovery of 80% or more when backpulsing and cleaning were used
  - Had flux declines resembling similar simulants with higher solids concentrations (factors of 10-20)
- A crossflow visualization cell was conceived to further study the topic of interactions leading to fouling at low solids concentration
  - Conceptual demonstration of viability as an experimental technique
  - Future work to investigate surface interactions that control fouling behavior

17 31 October 2017

Acknowledgements

Pacific Northwest NATIONAL LABORATORY
Proudly Operated by Battelle Since 1965

- Pacific Northwest National Laboratory
  - Justin Billing
  - Rick Shimskey
  - Ernest Antonio
  - Margaret Smoot
  - Renee Russell
  - Amy Rovira
- Washington River Protection Solutions
  - Matt Landon
  - Tom Sackett
- Savannah River National Laboratory
  - Dave Herman
- Howard University

18 31 October 2017

BACKUP SLIDES

Pacific Northwest NATIONAL LABORATORY
Proudly Operated by Battelle Since 1965

19 31 October 2017

CFF System Description

Pacific Northwest NATIONAL LABORATORY
Proudly Operated by Battelle Since 1965

- Single, 8-ft tubular filter element
- 25L slurry reservoir
- Rotary lobe progressive cavity pump
- Slurry recirculation loop
  - Flow instrumentation
  - Heating/cooling
  - Pressure instrumentation
- Permeate collection system
  - Flow instrumentation
  - Backpulse system
  - Backflush system

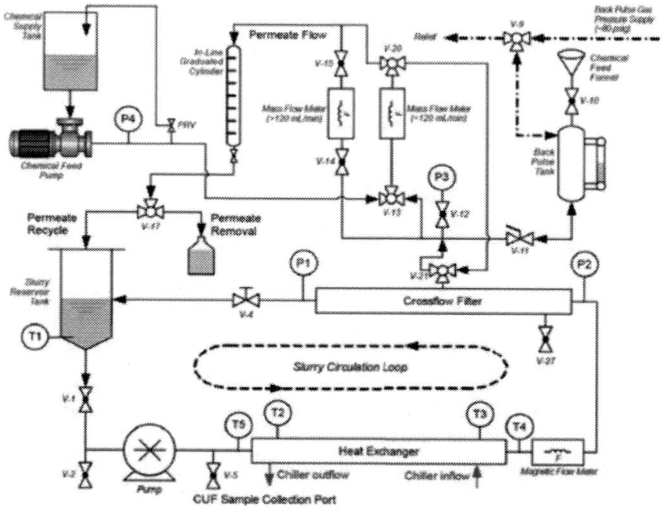

20 31 October 2017

CFF Test Operation [Typical]

Pacific Northwest NATIONAL LABORATORY
Proudly Operated by Battelle Since 1965

- Tests were conducted with permeate recycled back to reservoir
- Baseline conditions: 20°C, transmembrane pressure (TMP) of 20 psid, and an axial velocity (AV) of 14.7 ft s^{-1}.

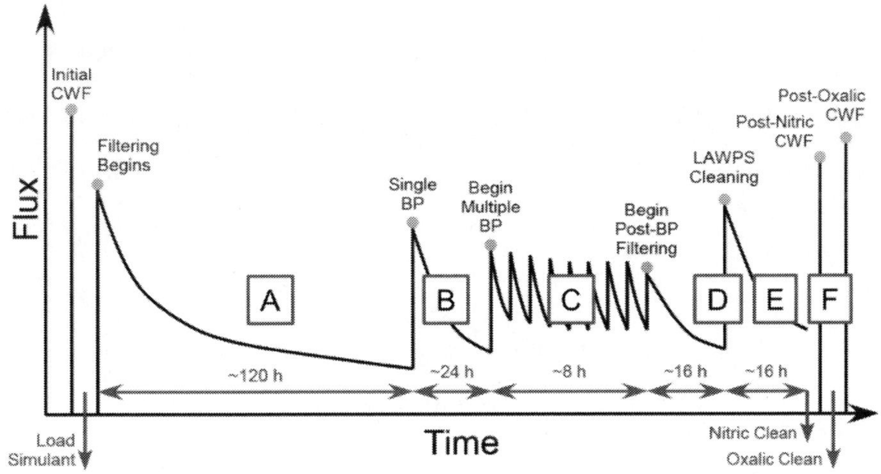

21 31 October 2017

CFF LAWPS Cleaning Approach

Pacific Northwest NATIONAL LABORATORY
Proudly Operated by Battelle Since 1965

- Traditional cleaning approach
  - Drain or displace slurry from system
  - Rinse with water, then acid (2M nitric, 0.5M oxalic) cleaning
- LAWPS cleaning protocol
  - Use ion exchange elution solutions to backflush the filter (shell to tube)
  - At bench-scale, preserve processing time by scaling the flow rate by the difference in filter area

LAWPS Step	Solution	Flow Rate (gpm [mL/min])	Volume (gal)	Time (hour)
(1) Displace	0.1 M NaOH	0.030 [114.6]	3.50	1.9
(2) Rinse	Process Water	0.030 [114.6]	2.33	1.3
(3) Elution	0.45 M HNO_3	0.014 [53.5]	9.08[a]	10.7
(4) Rinse	Process Water	0.014 [53.5]	3.03	3.6
Total Time				17.5

[a] Represents delivery of 50% of planned elution volume. The first 50% (approximately) is planned to be sent directly to waste.

22 31 October 2017

LAWPS 5.6M Simulant Basis

Pacific Northwest NATIONAL LABORATORY
Proudly Operated by Battelle Since 1965

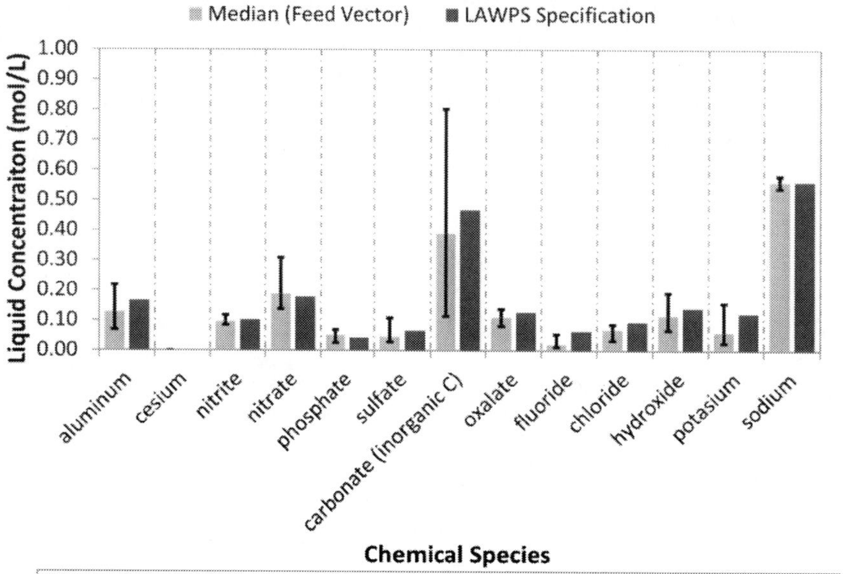

5.6M Liquid Concentrations

Dissolved Species	Nominal Na Simulant Concentration (mol L^{-1})
Al	1.66×10^{-1}
Cs	1.04×10^{-4}
Nitrate	1.78×10^{0}
Nitrite	1.02×10^{0}
Phosphate	4.32×10^{-2}
Sulfate	6.61×10^{-2}
Inorganic C	4.67×10^{-1}
Oxalate	1.27×10^{-2}
Chloride	1.22×10^{-1}
Free OH	1.41×10^{0}
Potassium	1.22×10^{-1}
Sodium	5.60×10^{0}

NOTE:
Sodium, Nitrite, Nitrate, Free OH concentrations all divided by factor of 10.
Oxalate concentration multiplied by factor of 10.

5.6M Na Simulant vs. Historical

Pacific Northwest NATIONAL LABORATORY
Proudly Operated by Battelle Since 1965

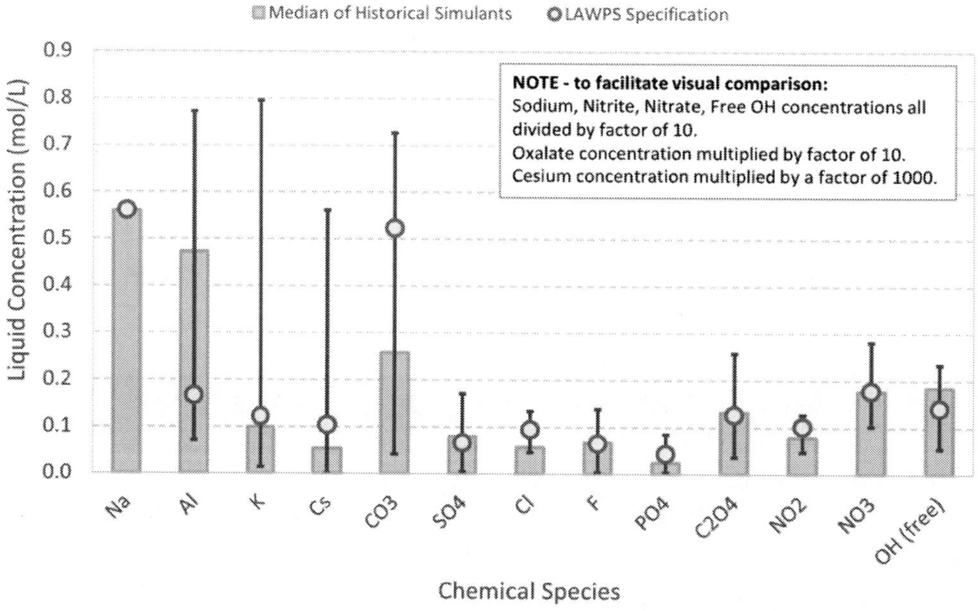

CFF Results: All Na Simulants, 20°C

Pacific Northwest NATIONAL LABORATORY
Proudly Operated by Battelle Since 1965

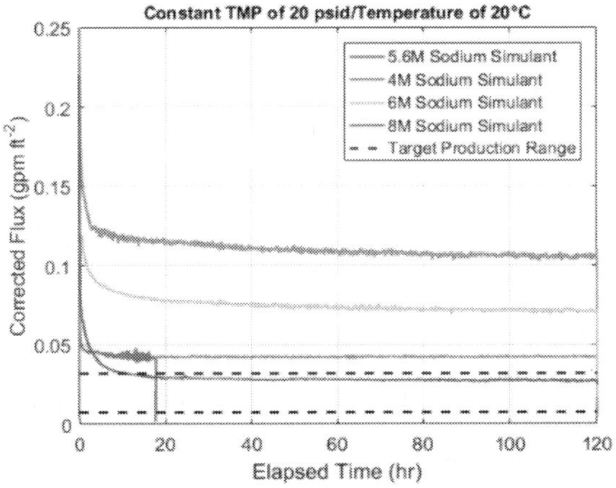

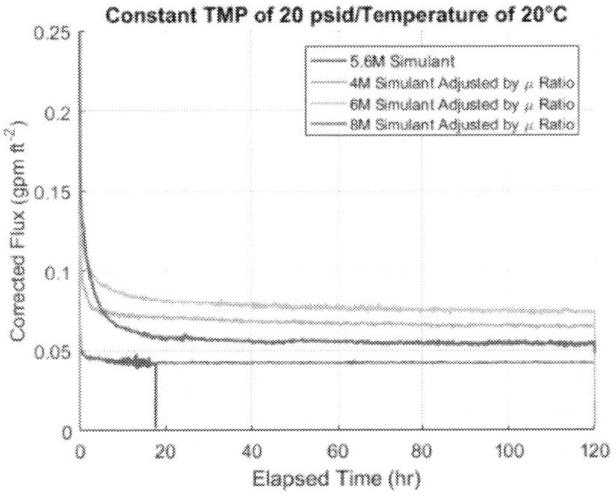

These simulants filtered as if solids were forming stable cake layer, with differences mostly accounted for by viscosity.

The target flux was achieve in every case, and psuedo steady-state was readily achieved.

CFF Historical Data Comparison

Pacific Northwest NATIONAL LABORATORY
Proudly Operated by Battelle Since 1965

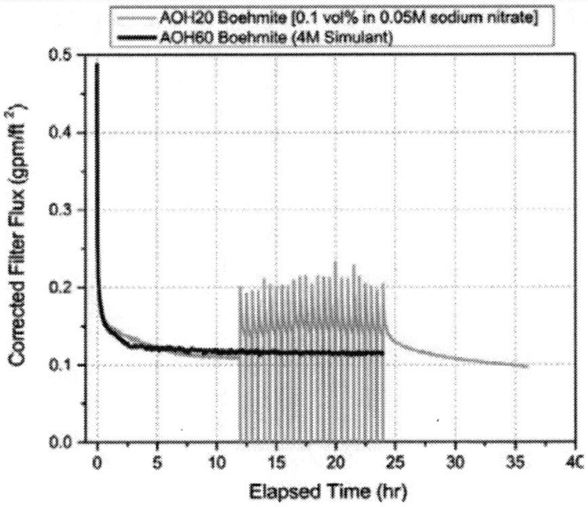

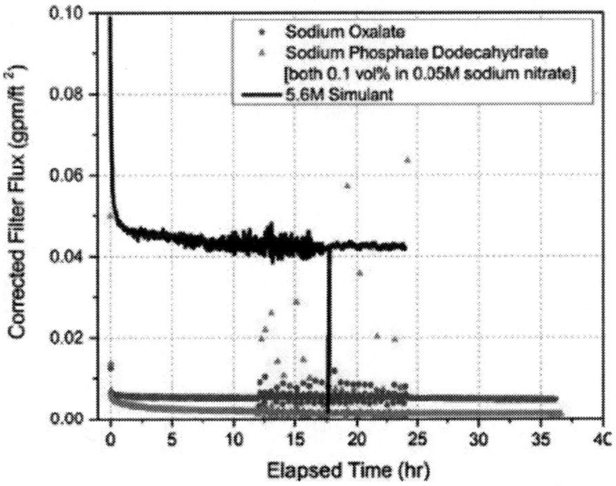

- All data corrected to a TMP of 20 psid, temperature of 25°C, and adjusted for differences in viscosity to be consistent with current CFF data.
- Current 5.6M simulant flux much higher than previous data.
- Boehmite filtered very similarly even though different types were used.

497988 Performance-Based Simulants for Hanford Radioactive Waste Treatment Process Testing

Abstract:

Radioactive wastes from nuclear fuel processing are stored in large underground storage tanks at the Hanford Nuclear Reservation near Richland, Washington. Treatment and remediation requires that waste feed from the storage tanks be delivered to the Hanford Tank Waste Treatment and Immobilization Plant (WTP) in a manner that ensures continuous WTP operations over the life cycle of the treatment mission. The complex physical and chemical processes required for this mission, the significant scale, and the hazardous nature of the waste necessitate the use of simulants in lab- and pilot-scale process testing.

As pointed out in the National Research Council (NRC) report *Advice on the Department of Energy's Cleanup Technology Roadmap: Gaps and Bridges*, testing with a complete range of actual radioactive wastes to cover all process limits related to chemical, physical, and rheological properties would require large amounts of actual waste that is difficult to retrieve and handle, increases radiological safety hazards to laboratory staff, and in most cases is too time-consuming and costly. The benefits of developing accurate and reliable waste slurry simulants are discussed in the NRC report:

"The absence of adequate understanding of the behavior of process streams can necessitate overly conservative and costly process designs to minimize the risk of a process failure or the risk of unrecognized safety issues, which as a worst case can render a facility inoperable with the actual radioactive waste it was intended to process."

Hanford simulants were initially developed by attempting to mimic the chemical composition of the waste. This was typically done by characterizing the chemical analytes of a tank waste sample, then matching the chemical composition either with precipitated metal hydroxides or by adding metal oxides. However, this approach can have significant drawbacks wherein the simulants developed that do not reproduce properties of the tank waste significant to treatment process performance, such as particle size or rheological properties. Because of these limitations, a new class of simulants, "performance-based simulants," was developed to match physical properties salient to waste processing step performance. This approach targets the performance of the simulant relative to actual waste performance data, and has been developed into ASTM standard C1750-1 for simulant development. Development and testing results of three specific Hanford waste performance-based simulants addressing the waste feed delivery system and leaching and cross-flow filtration performance in the WTP are presented. These simulants developed using specific Hanford waste performance characteristics and have been used to successfully demonstrate and understand significant aspects of process performance.

485286 Solids in Hanford's Low Activity Waste Pretreatment System Feed Tanks and Implications for Design Requirements

Abstract:

The Hanford Site has approximately 56 million gallons of nuclear waste generated from plutonium production. Much of this waste is highly concentrated electrolyte solutions of sodium-bearing salts that contain cesium-137 as the dominant radionuclide. The Hanford site is currently pursuing a Direct-Feed Low-Activity Waste treatment process (DFLAW). In the DFLAW process, any suspended solids are removed by filtration and the cesium is removed by ion-exchange in the Low-Activity Waste Pretreatment System (LAWPS), prior to being transferred to the Low Activity Waste Vitrification facility for immobilization. The vitrified glass product will be disposed of as low-activity waste.

A number of larger tanks (1.25 million gallon capacity) are used to prepare feed to the LAWPS. These large tanks contain accumulated solids in the bottom. The presence of solids in the tanks leads to a number of questions that must be answered in order to most efficiently feed LAWPS. The questions answered here are: What is the quantity of solids? What is the identity of the solids? How does the identity of the solids impact the filter cleaning strategy and the strategy of using temperature control to prevent salt precipitation? How do you minimize the amount of solids sent to LAWPS in order to maximize the throughput through the filtration system?

The quantity of solids in each tank was recorded using sludge weight measurements in each of the tanks, with some tanks having up to 112,000 gallons of solids. Even the feed tank (Tank 241-AP-107) had solids in the bottom. The identity of the solids was determined in waste samples by X-ray diffraction, polarized light microscopy, and scanning electron microscopy with energy dispersive spectroscopy. All tanks were found to have thermonatrite ($Na_2CO_3*H_2O$), sodium oxalate ($Na_2C_2O_4$), kogarkoite (Na_3FSO_4) and natrophosphate ($Na_7F(PO_4)_2*19H_2O$). All of these phases were salts, so it can be concluded the waste will be saturated with these salts prior to dilution. The Hanford Tank Farm Contractor intends to dilute these wastes prior to feeding them to the LAWPS, but if care is not taken during dilution, additional salt can dissolve, re-saturating the dilute supernate with these salts. Thus, the mixing process should be designed to minimize the mixing of the dilute supernate with the solids layer. The LAWPS feed tank, which is continuously recirculated during operation, should maintain the pump inlets and outlets as far away from the solids layer as possible while still maintaining sufficient volume in the tank for processing so that currents within the tank do not interact with the solids layer.

The solubility of three of the four salts increases with temperature, so their precipitation can be prevented in the LAWPS facility by slightly increasing the temperature of the incoming feed by 1 to 2 °C. The solubility of thermonatrite, however, decreases with increasing temperature, which means this salt could precipitate if it is near saturation and the temperature is increased. Nonetheless, a small temperature increase should not cause thermonatrite precipitation as long as the sodium molarity of the feed to LAWPS is maintained below 6 M; a value determined by evaluating the solubility of thermonatrite in aqueous solutions of the three most prevalent electrolytes in Hanford waste ($NaOH$, $NaNO_2$, and $NaNO_3$).

498264 Hazard Analysis Under Nuclear Regulations

Abstract:

Hazard analysis with its 4 basic steps of hazard identification, event postulation, control identification, control selection with an overarching safety risk impact is the building block of safety evaluation and management. The presentation will build on the hazard analysis process for chemical and overlay the regulatory and process impact associated with nuclear hazards. The Department of Energy (DOE) regulations, standards, handbooks, and guides will be discussed rather than the requirements from the Nuclear Regulatory Commission. The presentation will show similarities in meeting EPA and OSHA requirements along with other DOE requirements. A discussion of the DOE approved codes tools box used for chemical consequence evaluation along with evaluation guidelines.

500137 Model-Predictive Safety System for Predictive Detection of Operation Hazards

Abstract:

In 2016 we introduced a method of model-predictive safety (MPS) system design [1]. The MPS system generates alarm signals that are predictive and systematically account for process nonlinearities and interactions, while typical existing functional safety systems generate reactive, non-interacting alarm signal(s) when a process variable exceeds a threshold, For the first time, the MPS system design method proposed a systematic utilization of the dynamic process models to generate predictive alarm signals (alerts) for the detection of present and future operation hazards (OHs) in real time. An MPS system uses a process model to project in real-time the process operability status and to generate alarm signal(s) indicating the presence of a present or future OH with reasonable accuracy; it generates predictive alarm signals that alert the process personnel to imminent and potential future OHs before the actual OHs occur. Although the method uses the concepts of moving-horizon, model-based prediction and state estimation, it does not implement any control; when an MPS system identifies a current or future hazard in the operation of a process, it indicates that no controller is or will be able to assure safe process operation. Compared to a first-principles MPC system, an MPS system is easier to implement, and its on-line computational cost is significantly less.

In this paper, we present results from the application of the MPS system design method to chemical process examples. The advantages of the predictive and interacting nature of the alarm signals that the MPS systems generate over conventional alarm signals are shown.

References

[1] Mohseni Ahooyi, T., J.E. Arbogast, W.D. Seider, U.G. Oktem, and M. Soroush, "Model-Predictive Safety System for Proactive Detection of Operation Hazards," AIChE J., 62, 2024-2042 (2016).

RADIOLOGICAL IMPLICATIONS OF THE PHYSICOCHEMICAL FORMS OF STRONTIUM RELEASED DURING POSTULATED NUCLEAR WASTE FACILITY ACCIDENT CONDITIONS

Maeley K. Brown, David C. Thoman, and Kevin R. O'Kula
AECOM N&E Technical Services, LLC
Aiken, South Carolina
maeley.brown@aecom.com; dave.thoman@aecom.com; kevin.okula@aecom.com

Abstract

In preparing Documented Safety Analyses (DSA) for U.S. Department of Energy (DOE) Complex facilities, DOE-STD-3009-2014 is the governing accident analysis standard for compliance with DSA requirements of 10 C.F.R Part 830, the Code of Federal Regulations pertaining to nuclear safety management. The standard provides three options for the evaluation of atmospheric dispersion and relative concentration (χ/Q) for use in estimating radiological doses for postulated accident scenarios. These options, while serving the same purpose of estimating radiological doses for postulated accident scenarios, implement different assumptions and inputs, thus yielding varying results.

This paper compares the radiological dose estimates of all three options for unit-activity releases of a radionuclide, Strontium-90 (Sr-90), found in many waste streams from nuclear processes, and quantifies the respective level of conservatism for each option. Option 1 calculations are performed within an Excel workbook. Options 2 and 3 are evaluated using a DOE Safety Software Central Registry atmospheric transport and dispersion (ATD) modelling computer software, with the Option 3 calculations making use of site-specific surface roughness and deposition velocity input data. This analysis further seeks to identify the causes of the different results provided by the above three options. In addition, the effect of the physicochemical form of strontium on dose is quantified.

Introduction

In 2014, the Department of Energy Standard 3009 (DOE-STD-3009) was fully revised.[1] This revision includes additional guidance on the calculation of radiological consequence. DOE-STD-3009-2014 describes Radiological Dose Consequence in section 3.2.4.2 to be calculable in three different ways. In this paper, these three methods of analysis are calculated and compared to one another for a postulated point source release of Sr-90. This paper primarily analyzes the calculated impact to the general public based on such a release. The three calculation methods in DOE-STD-3009-2014 are as follows:

- Option 1: Follow a process based on Nuclear Regulatory Commission (NRC) Regulatory Guide (RG) 1.145,[2] Atmospheric Dispersion Models for Potential Accident Consequence Assessments at Nuclear Power Plants;
- Option 2: Use a DOE-approved toolbox code and apply the conservative parameters defined in DOE-STD-3009-2014;
- Option 3: Use site-specific methods and parameters as defined in a site/facility-specific DOE-approved modeling protocol.

- In this paper, DOE Central Registry ATD software MELCOR Accident Consequence Code System 2 (MACCS2), version (1.13.1) is applied to perform calculations under Options 2 and 3. [3]

The three options presented in DOE-STD-3009-2014 for the radiological dose calculation are primarily based on different methods for determining the atmospheric relative concentration (χ/Q) that represents the normalized dilution of the radioactive material in the plume as it travels from the source location to the downwind location of the maximally exposed offsite individual (MOI) of the general public. During transport, dilution occurs through various mechanisms that include effects for dispersion through atmospheric turbulence and depletion through deposition. The horizontal and vertical dispersion coefficients (σ_y and σ_z, respectively) are key inputs to the χ/Q calculation, along with the wind speed, and represent the extent of the plume growth as a function of downwind travel distance. Larger dispersion coefficients are associated with more unstable atmospheric conditions and lead to lower χ/Q values. Atmospheric stability is typically designated by Stability Classes A (extremely unstable) through either F (moderately stable) or G (extremely stable) and is inferred from collected meteorological data on an hourly basis.[4,5] DOE-STD-3009-2014 requires a set of meteorological data which is recent and representative of five years for the calculation of offsite dose.

Option 1

The first calculation option is to use the methodology described in the NRC RG 1.145. Here, guidance is provided to manually calculate the appropriate dispersion coefficients and to develop a useable database of meteorological data, based on NRC RG 1.23.[5]

The meteorological data used in this paper was collected by the Vogtle Electric Generating Plant (Plant Vogtle), and is collected on an hourly basis. This data is considered to be representative of the Savannah River Site (SRS) based on its proximity of fewer than 20 miles from the site boundary.

Using Option 1, dose consequence is calculated manually, typically with the use of a spreadsheet because of the large number of repetitive calculations. NRC RG 1.145 provides graphs from which to calculate σ_y, σ_z, and plume meander factor, which are used in addition to wind speed and wind direction, provided from the input meteorological data. Based on these factors, the χ/Q is calculated and applied to the TED equation.

Option 2

The second calculation option is to use a DOE pre-approved toolbox code with conservative input parameters, provided in the DOE-STD-3009-2014, applied. The DOE Office of Environment, Health, Safety and Security maintains a list of software that has been evaluated against the department's Safety Software Quality Assurance requirements and related safety software guidance. This list of qualified software is deemed the Central

Registry, or "toolbox" software. This computer software is used by DOE contractors to perform calculations used to establish the safety basis for DOE nuclear facilities and their operation, and to support the variety of safety analyses and safety evaluations developed for these facilities. Included in the set of toolbox software is MACCS2, a Gaussian plume segment model for calculation of radiological atmospheric dispersion and consequences due to postulated atmospheric releases of radionuclides from nuclear facilities.[6]

MACCS2 is developed and maintained by Sandia National Laboratories (SNL). It is the chief consequence code used in probabilistic safety analysis for numerous commercial nuclear power plants throughout the world. In the DOE complex, it is the most applied software tool for deterministic consequence analysis. Specifically, use of MACCS2 in DOE applications is primarily to support decision-making on control selection mostly in the identification of safety structures, systems, and components (SSCs). The Central Registry-recognized version is 1.13.1 from 2004,* therefore this is the version applied for modeling a postulated Sr-90 release using Options 2 and 3 as described above.

For most DSA applications, MACCS2 use follows stratified random sampling of qualified, site meteorological data, where each hour of the year's data is analyzed. Also, of importance in the present application, the deposition velocity of Sr-90 and the surface roughness of the SRS are user-specified as explicit parameter inputs.

MACCS2 is divided into three modules. The module in which the radioactive dilution calculations are performed is the ATMOS module. The ATMOS module calculates atmospheric transport, dispersion, deposition, and radioactive decay. This module stores the information on wind direction, arrival and departure times, and plume dimensions. The EARLY module performs emergency phase calculations. This module stores information on the emergency such as duration, exposure pathway, and mitigative actions. The CHRONC module pertains to the intermediate and long-term release phases, using mitigative action models.[3]

The inputs applied to MACCS2 for Option 2 are as follows:

- Non-buoyant, ground-level, point-source release;
- Plume centerline concentrations;
- Rural dispersion coefficient;
- DOE-STD-3009-2014 specified deposition velocity of 0.1 cm/s for unfiltered particulate release in a respirable range of 1-10 μm Aerodynamic Equivalent Diameter (AED);
- DOE-STD-3009-2014 specified surface roughness of 3 cm;
- Minimum wind speed of 0.5 m/s, as this is the minimum fixed value within MACCS2;
- Averaging Time Plume Meander Model;
- Building wake factors are not credited.

---

* The current version of MACCS2 is 3.9 (http://maccs.sandia.gov/events.aspx), released in 2014.

Additional considerations include:

- The plume is modeled for one day (24 hours) post-release;
- Precipitation effects are not considered in this analysis;
- Because inhalation is the pathway considered in these TED calculations, food and water ingestion pathways are not modeled.

Option 3

Option 3 is unique in that it does not require a specific method of calculation. However, sites which choose to implement this option typically utilize a DOE toolbox code with site-specific parameters for speed and precision. Due to this trend, this paper uses MACCS2 and with site-specific input parameters. This paper utilizes parameters specific to the SRS, which is populated with forests interspersed throughout with industrial areas. This topography allows for a greater surface roughness and deposition velocity than the conservative DOE-STD-3009-2014 specified values in Option 2. This is expected to result in slightly lower χ/Q values as a result of the more realistic, less conservative parameters. A summary of all three modelling options is presented in Table 1.

Table 1: Summary of Key Model Attributes

Model Attribute	Option 1 (NRC RG 1.145)	Option 2/3 (MACCS2)		Comment
Stability Class	A through G	A through F		NRC RG 1.145 allows for the very stable class G. However, MACCS2 converts G values to F stability class.
Dispersion Coefficients	Eimutis-Konicek (E-K)	Eimutis-Konicek (E-K)		The Eimutis-Konicek dispersion model supports G stability class and is based on a 3 cm surface roughness and 3 minute averaging time.
Source Roughness	3 cm	Option 2: 3 cm	Option 3: 160 cm	NRC RG 1.145 methodology does not use an algorithm for adjusting σ_z for surface roughness.
Plume Meander Model	NRC Model	Averaging Time Model		The NRC model is the required model for Option 1. Three minutes is used as the averaging time for Options 2 and 3 baseline cases.
Deposition Velocity	Not Applicable	Option 2: 0.10 cm/s	Option 3: 0.61 cm/s	Options 2 and 3 allow for plume depletion from deposition for particulate releases, but this is not an option with the NRC RG 1.145 methodology.
Minimum Wind Speed	0.1 m/s	0.5 m/s		The wind speed in the Plant Vogtle meteorological data file has a minimum wind speed of 0.1 m/s. MACCS2 has a minimum wind speed limit of 0.5 m/s. Where the meteorological data is less than 0.5 m/s, MACCS2 assumes the speed is 0.5 m/s.

Radiological Information for Strontium

To quantify the dose consequence of an event, the Total Effective Dose (TED) equation takes into account inhalation and direct exposure to a release. In this analysis, a release of Sr-90 is modelled under each of the three DOE-STD-3009-2014 options. Compounds of Sr-90 (a radioactive half-life of 28.79 years) are present in nuclear waste throughout the DOE Complex.[7] Because Sr-90 is primarily a source of beta-particle radiation and direct exposure from cloudshine and groundshine is negligible, inhalation dose is the dominant pathway in calculating the dose consequence to the general public. Inhalation dose is calculated from the mass of material released (Source Term (ST)), the concentration of the material in the air (χ/Q), the Dose Conversion Factor (DCF), and the Breathing Rate (BR) of the hypothetical impacted individual.

DCFs are specific to individual radionuclides and a radionuclide can have more than one based on the manner in which it is absorbed into the body. DOE-STD-3009-2014 specifies that dose coefficients consistent with International Commission on Radiological Protection Publication 72, *Age-dependent Doses to Members of the Public from Intake of Radionuclides* shall be used for public dose estimates.[8] ICRP Publication 72 incorporates the recommendations from ICRP Publication 71 which notes that the physicochemical form and particle size of a radionuclide are incorporated in a dosimetric model that quantifies retention of deposited activity in the respiratory tract regions and clearance to other organs by representing clearance processes and ultimately absorption of radiological material into body fluids.[9] Sr-90 has three possible absorption types: Fast (F), Moderate (M), and Slow (S). A Fast absorption type is readily absorbed into the blood from the respiratory tract while a Slow absorption type is relatively insoluble. Experimental and accident evidential data show mostly moderate and fast types, with the Slow type appearing mostly for cases where a compound such as strontium titanate is released. In absence of specific information, ICRP 71 recommends that type M be used as a default basis. It should be noted that ICRP 68 and ICRP 119 provide a recommendation of type S for strontium titanate and an alternative recommendation for type F for the effect of unspecified compounds to the worker receptor.[10,11] It is noted that strontium titanate is not commonly found in nuclear waste inventories.

In addition to the absorption type, particle size is also a factor in the determination of the appropriate DCF. For a release to the public, a 1 micrometer particle diameter is assumed to be conservatively representative. Table 2 lists the DCFs which are applied to the postulated Sr-90 release and are found in ICRP 119, 71, and 72.

<u>Table 2: Dose Conversion Factors of Sr-90 Absorption Types</u>

Absorption Type	DCF (Sv/Bq)
Fast	2.4E-08
Moderate	3.6E-08
Slow	1.6E-07

Results

Figure 1 compares the analysis Options 1, 2, and 3 based on the moderate absorption type of strontium at distances of one to ten km from the point of release. Over this distance, it is clear that the Option 3 analysis method yields the lowest TED, and thus the least conservative results of the three options. It can be easily seen from this figure that Option 1 becomes increasingly more conservative, when compared to Options 2 and 3, as the distance from the point of release increases.

The Option 1 TED result is comparable to the Option 2 TED result at one km due the NRC plume meander factor which has the largest impact in lowering the dose close to the source and which serves to counterbalance the conservatisms noted below associated with this option. Given that this effect is only applied for plume travel distance up to 800 m, the effect becomes increasing less pronounced beyond 800 m. At larger distances, the Option 1 doses are considerably higher in comparison to those of Options 2 and 3 due to the use of a lower minimum wind speed and consideration of G stability class.

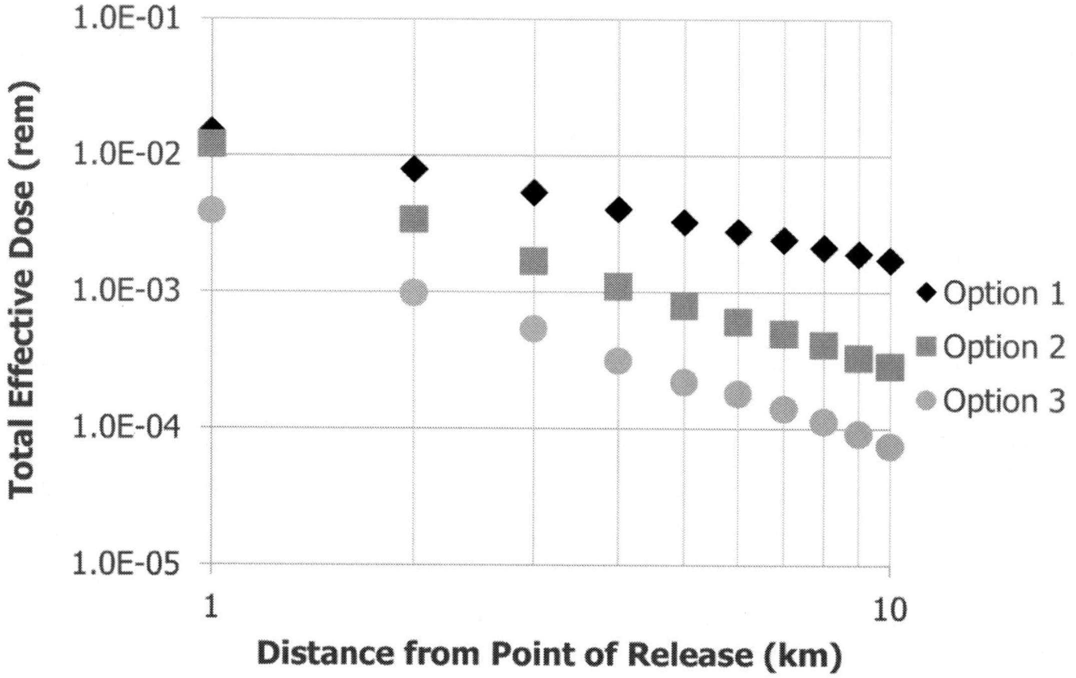

Figure 1: Comparison of Radiological Consequence Calculation Options

Regulatory guidance in DOE-STD-3009-2014 states that appropriate χ/Q values at the general public receptor shall be determined using a method consistent with application of Regulatory Guide 1.145, using either the directionally independent or directionally dependent method. For directionally independent assessments, this calculation represents the 95th percentile result regardless of wind direction, as described in Regulatory Guide 1.145, Section C.3, Regulatory Position 3 (Determination of 5 Percent Overall Site χ/Q Value). Directionally-dependent calculations are performed separately for each wind direction sector and the result represents the maximum 99.5th percentile among the sectors, as described in Regulatory Guide 1.145, Section C.2, Regulatory Position 2 (Determination of the Maximum Sector Values).

Figure 2 compares the directionally independent and dependent analysis results for the Option 1 analysis method based on a moderate strontium absorption type. The values are within 10-15% agreement at each distance, however the directionally independent method consistently yields slightly greater, and thus slightly more conservative, TED.

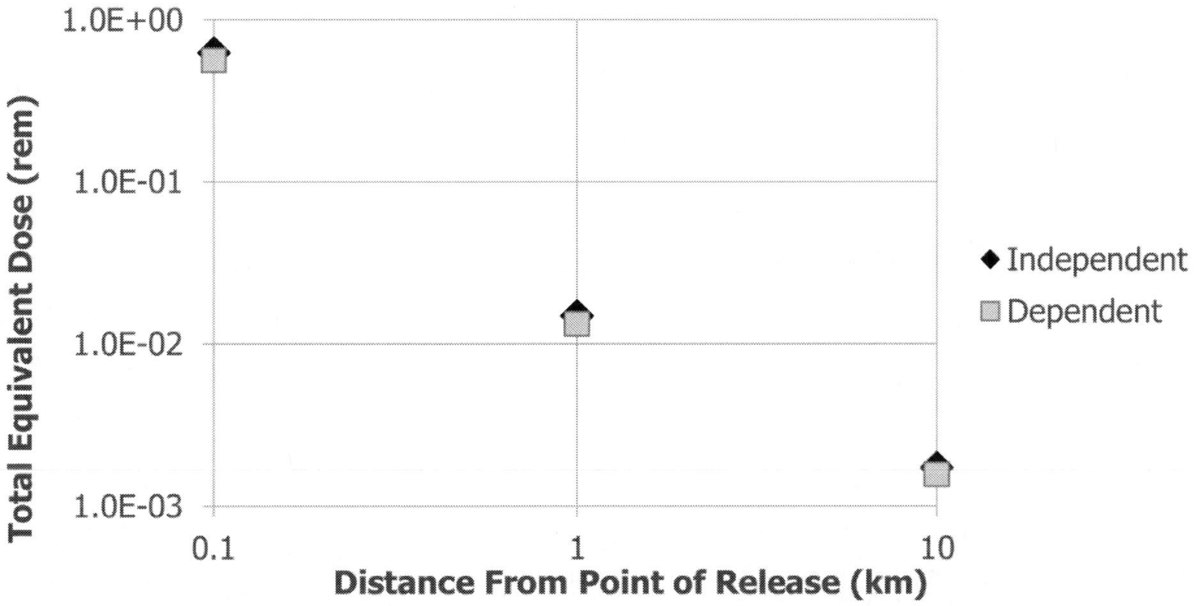

Figure 2: Comparison of Directionally Independent and Dependent Analyses

Figure 3 compares the TED of the Sr-90 absorption types to the general public based on an Option 2 analysis method. It can be clearly perceived that the slow absorption type yields the greatest, and thus the most conservative, TED values for each analyzed distance from the point of release. The fast absorption type consistently yields the lowest results. The ratios between each calculated TED are is directly proportional to the ratios between the DCF of each absorption type. The moderate type DCF and TED are 23% of the slow type, and the fast type DCF and TED are 67% of the moderate type.

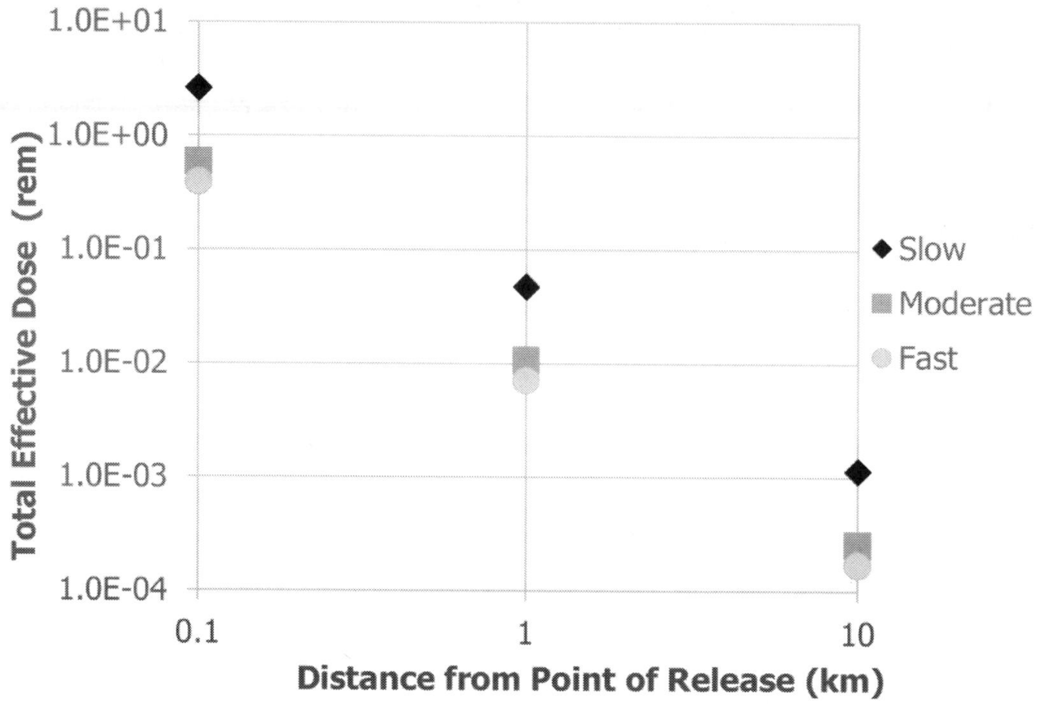

Figure 3: Comparison of Sr-90 Doses Based on Assumed Absorption Type Released

Conclusions

The 2014 revision of the DOE-STD-3009 standard provides radiological consequence calculation guidance and in turn, allows for increased flexibility for the dispersion analyst in the three options provided, with Option 1 allowing the least user flexibility and Option 3 the most. At distances of 1 km and greater from the point of release, Option 1 yields the highest, and thus most conservative, predictions. At these distances, Option 3, containing the most site-specific input data, yields the lowest, and thus least conservative, predictions. Site-specific topographical data significantly decreased the conservatism of the dispersion model, as represented by a comparison of Option 2, conservative inputs, and Option 3, site-specific inputs.

The distribution of calculated χ/Q values can be analyzed either in a directionally independent or directionally dependent manner. In this application using Plant Vogtle meteorological data, the consequences calculated with the directionally independent method are 10-15% greater than consequences calculated from the directionally dependent method for all distances evaluated.

Sr-90 can be released as fast, moderate, and slow absorption type compounds under postulated accident conditions. Applying the Slow absorption rate to the postulated release (e.g., strontium titanate) yields the highest, and thus the most conservative, consequence result.

References

1. "Preparation of Nonreactor Nuclear Facility Documented Safety Analysis," DOE-STD-3009-2014, Washington, DC: U. S. Department of Energy, 2014.
2. "Atmospheric Dispersion Models for Potential Accident Consequence Assessments at Nuclear Power Plants," Regulatory Guide 1.145, Washington, DC: U. S. Nuclear Regulatory Commission, 1983.
3. Chanin, D., M. L. Young, J. Randall, and K. Jamali, *Code Manual for MACCS2: Volume 1, User's Guide*, NUREG/CR-6613, Volume 1, U. S. Nuclear Regulatory Commission, Washington, D. C., May 1998.
4. *Meteorological Monitoring Guidance Regulatory Modeling Applications*, EPA-454/R-99-005, Research Triangle Park, NC: U. S. Environmental Protection Agency, 2000.
5. "Meteorological Monitoring Programs for Nuclear Power Plants," Regulatory Guide 1.23, Washington, DC: U. S. Nuclear Regulatory Commission, 2007.
6. "MACCS2 Computer Code Application Guidance for Documented Safety Analysis," DOE-EH-4.2.1.4-MACCS2-Code Guidance, Washington, DC: U. S. Department of Energy, 2004.
7. *Derived Concentration Technical Standard*, DOE-STD-1196-2011, U.S. Department of Energy, Washington, DC, May 2011.
8. International Commission on Radiological Protection (ICRP). Age-Dependent Doses to Members of the Public from Intake of Radionuclides, Part 5. Compilation of Ingestion and Inhalation Dose Coefficients. ICRP Publication 72, Pergamon Press, Oxford, 1996.
9. International Commission on Radiological Protection (ICRP). Age-Dependent Doses to Members of the Public from Intake of Radionuclides, Part 4. ICRP Publication 71, Pergamon Press, Oxford, 1995.
10. International Commission on Radiological Protection (ICRP). Dose Coefficients for Intakes of Radionuclides by Workers. ICRP Publication 68, Pergamon Press, Oxford, 1994.
11. International Commission on Radiological Protection (ICRP). 2012. Compendium of Dose Coefficients based on ICRP Publication 60. ICRP Publication 119. Ann. ICRP 41(Suppl.) Oxford, UK Oxford, UK: Elsevier, Elsevier, 2012.

Assessing Consequences of Chemical Releases on Control Room Habitability using HABIT 2.1

Thomas O. Spicer, III
Ralph E. Martin Department of Chemical Engineering
University of Arkansas
Fayetteville, AR 72701

Casper Sun
Syed Haider
U.S. Nuclear Regulatory Commission
Rockville, MD 20852

HABIT is a package of computer codes designed to assist in the evaluation of control room habitability in the event of accidental spills of hazardous materials in the area and is part of the U.S. Nuclear Regulatory Agency's Radiological Protection Computer Code Analysis and Maintenance Program (RAMP). Developed at Pacific Northwest Laboratory (PNL), HABIT was originally written to model atmospheric dispersion based on standard passive dispersion models. In addition to an extensive update of the user interface, HABIT 2.1 was also extended to include the effects of denser-than-air contaminant cloud behavior using the SLAB and DEGADIS models. This paper summarizes how the HABIT source models (instantaneous and finite duration releases) were used to generate inputs to SLAB and DEGADIS. Model results are presented for a hypothetical chlorine release, and the importance of denser-than-air behavior is discussed including the specification of the air intake height on the impact to a control room.

1. Introduction

HABIT 2.1 is a collection of computer models designed to assist in the evaluation of control room habitability in the event of an accidental spill of toxic industrial chemicals (TICS) or radionuclides. Past versions of HABIT (Ramsdell, 1991) produced tabular files that could be imported into spreadsheet programs for further analysis, and (passive) dispersion calculations were made using EXTRAN. In addition to graphical output, HABIT V2.1 also incorporates DEGADIS (Spicer and Havens, 1989) and SLAB (Ermak, 1990) dispersion models addressing denser-than-air behavior of many important TICS. Criteria suggested by Bitter and McQuaid (1988) are used to recommend when the denser-than-air behavior should be important to consider. EXTRAN remains unchanged from previous versions of HABIT.

2. Required Model Inputs

Model inputs for HABIT are similar to what is expected for typical atmospheric dispersion models:

Meteorological conditions:

- Atmospheric stability class

- Wind speed at 10 m, U_a
- Air temperature, T_a
- Atmospheric pressure, P_a
- Solar radiation, R_s
- Cloud cover (tenths)
- Surface roughness, z_R
- Relative humidity

Chemical contaminant properties:

- Molecular weight, M_c
- Normal boiling point, T_b
- Liquid heat capacity, C_p
- Heat of vaporization, H_v
- Specific gravity (liquid), ρ_L
- Vapor phase molecular diffusivity, D
- Vapor phase heat capacity, C_{pv}

Spill input parameters

- Initial mass, m
- Mass release rate, m_r
- Release height above ground level, h
- Contaminant storage temperature, T_s
- Contaminant storage pressure, P_s
- Maximum pool radius, R_{max}
- Distance to intake, X_p
- Intake height, Z_p
- Building area (normal to the wind direction)
- Vent flow rate, F_v
- Averaging time

3. Aerosol Behavior

EXTRAN models flashing liquid releases assuming the liquid phase completely rains out on the ground, but this approach has been questioned. Witlox et al. (2010) considered data from several sources and concluded that there was a transition between mechanical break up and fully flashing jets for superheats (temperature above the boiling point) between 10 and 20°C. For the SLAB and DEGADIS calculations, a conservative assumption would be to assume that the release fully flashes to form an aerosol cloud when the superheat exceeds 10°C.

In DEGADIS and SLAB simulations of releases that could form an aerosol, the liquid is assumed to be retained in the dispersing cloud as an aerosol (typically considered a conservative assumption). SLAB models the condensed liquid phase directly. In DEGADIS, aerosol releases are typically modeled by specifying that the cloud density is a linear function of the contaminant concentration (at zero contaminant concentration, the density is that of air, and for the aerosol cloud, the initial concentration and density are equal before any dilution with air). This assumption was found to be valid based on adiabatic mixing calculations with (humid) air for ammonia and chlorine.

4. Britter and McQuaid Criteria for Recommending Use of a Denser-than-Air Dispersion Model

Britter and McQuaid proposed correlations for modeling the dispersion of denser-than-air contaminants from area sources for plumes and puffs based on field test data. Their analysis identified the dominant independent variables as:

- Density of released contaminant after depressurization to atmospheric pressure (ρ_s);
- Volumetric release rate (m_r/ρ_s) or total volume released (m/ρ_s);
- Characteristic wind speed (u_r); and
- Characteristic source dimension (D_s).

Based at least in part on the fact that available field test data for denser-than-air contaminants do not clearly indicate the importance of these parameters, Britter and McQuaid considered some independent parameters to be less important and consequently not included in their analysis (e.g., surface roughness, atmospheric stability, and exact source dimensions). The correlations proposed by Britter and McQuaid were also limited by the lowest concentrations that were observed (1000 to 2000 ppm). They also proposed criteria by which the importance of denser-than-air effects can be assessed. For steady-state plume releases, Britter and McQuaid found that denser-than-air effects are important if:

$$\frac{g(m_r/\rho_s)}{D_s u_r^3}\left(\frac{\rho_s-\rho_a}{\rho_a}\right) > 0.15^3 = 0.0034 \qquad (1)$$

where g is the acceleration due to gravity and ρ_a is the ambient air density which can be estimated with the ideal gas law. For puffs or instantaneous gas releases, denser-than-air effects are important if:

$$\frac{g(m/\rho_s)^{1/3}}{u_r^2}\left(\frac{\rho_s-\rho_a}{\rho_a}\right) > 0.2^2 = 0.04 \qquad (2)$$

The choice of values to be used in these criteria depend on the release type and how it is modeled in SLAB and DEGADIS. In all cases below, the characteristic wind speed is taken to be the ambient wind speed U_a used in EXTRAN.

5. Averaging Time

The averaging time (t_{av}) used in DEGADIS and SLAB accounts for the effect of plume meander and depends on the hazard under consideration (t_h) as well as the amount of time required for the release to become air borne (t_d). For flammable materials, t_h can be taken to be zero, and for toxic materials, t_h should be based on the criterion for comparison (e.g. TLV-STEL is based on an exposure duration of 15 min so t_h = 900 s). The length of time required for the release to become air borne depends on the spill scenario. DEGADIS and SLAB also have a lower limit of the averaging time (t_{min}) based on the model formulation (t_{min} = 20 s for DEGADIS and 10 s for SLAB). The averaging time input to the models should be taken as the max(min(t_h, t_d), t_{min}). The current implementation assumes that hazards under consideration are evaluated with the STEL, so t_{haz} = 900 s.

6. Release Scenarios Modeled by HABIT

HABIT (and EXTRAN) was developed to consider the hazards based on the following scenarios:

- Liquid Tank Burst
- Liquid Tank Leak
- Gas Tank Burst
- Gas Tank Leak

For these scenarios, EXTRAN estimates the rate at which material becomes airborne before the subsequent atmospheric dispersion calculations are made. DEGADIS and SLAB are both strictly atmospheric dispersion models in the sense that they require the airborne rate of the dispersing material, and the rate at which material becomes airborne must be provided by the user. For consistency between model predictions, the EXTRAN output was used to provide the rate the released material becomes airborne, particularly for evaporating liquid pools.

6.1. Liquid Tank Burst

In EXTRAN, this case generates a finite duration release based on evaporation from a spreading liquid pool. If the vapor pressure at the storage conditions is higher than ambient pressure, then there will be at least some flashing. EXTRAN models a flashing release by assuming all of the liquid phase rains out and subsequently evaporates from a liquid pool.

For releases conditions that produce an aerosol, DEGADIS and SLAB model this release case as an instantaneous release. The source area is set assuming the height-to-diameter ratio of the initial cloud is 0.5 assuming the cloud density is the undiluted aerosol density based on the (vapor) flash fraction f_v.

For release conditions that do not produce an aerosol, DEGADIS and SLAB use the source parameters calculated by EXTRAN to model the evaporating pool. In SLAB, there is no provision for a circular ground-level area source that changes diameter as a function of time, so the source area used in SLAB was taken to be the average of the area source predicted by EXTRAN. DEGADIS can take into account the initial flash fraction if present as an initial cloud over the evaporating (area source) pool. The initial flash fraction is taken from the EXTRAN output.

6.2. Liquid Tank Leak

In EXTRAN, this case generates a finite duration release based on evaporation from a spreading liquid pool for the duration of the release. As for the tank burst, if the vapor pressure at the storage conditions is higher than ambient pressure, then there will be some flashing.

For release conditions that produce an aerosol, SLAB simulates the release as if the flashing release occurred from a ground level, area source (A_s) as a vertical jet with velocity equal to the ambient wind speed (so $A_s = m_r/(U_a\rho_s)$). DEGADIS simulations are made using a ground-level area source with the same area source extent as used for SLAB.

In SLAB, the source area is the average of the area source predicted by EXTRAN as discussed above. DEGADIS uses the area source as a function of time as calculated by EXTRAN.

6.3. Gas Tank Burst

A gas tank burst might normally be simulated assuming some sort of initial dilution with air because of the explosive release, but EXTRAN does not make that assumption. For consistency with EXTRAN, the input file specifications for SLAB and DEGADIS do not include initial cloud dilution due to an explosive release.

In simulating this case with SLAB and DEGADIS, the gas undergoes an adiabatic expansion to ambient pressure, and this final state is used to calculate the source gas temperature T_s. The initial cloud is again assumed to have a height-to-diameter ratio of 0.5.

6.4. Gas Tank Leak

This scenario can be simulated with SLAB as a horizontal jet (in the same direction as the wind) or as a vertical jet. For DEGADIS, the jet can be vertical or the release can be simulated so that the release momentum is dissipated at the source without dilution; this would be a conservative approach which is consistent with the treatment outlined above for a gas tank burst. For this scenario, SLAB and DEGADIS estimate the area of the release assuming the vertical jet velocity is the same as the ambient wind speed as discussed above. The source gas temperature is estimated from adiabatic expansion of the gas from its storage conditions as discussed previously.

6.5. SLAB and DEGADIS input to HABIT

DEGADIS and SLAB estimate the concentration as a function of time for input to HABIT. The predicted concentrations from DEGADIS and SLAB should account for the effect of plume meander (reduction in the ensemble-average concentration due to anticipated wind direction variation), but the input to HABIT should not have other averaging effects (such as integrating the concentration time history at a point to obtain a time-averaged concentration). For both SLAB and DEGADIS, the chemical source location is assumed to be at ground level directly upwind of the control room intake. For both models, simulated concentration as a function of time is estimated at the intake height ($z = Z_p$) and distance from the release ($x = X_p$) on the plume centerline ($y = 0$).

In DEGADIS, the default model output summarizes estimated plume parameters as a function of distance at specified times. HABIT uses a post-simulation tool in DEGADIS to provide the needed model output as a function of time (file extension SR5).

In SLAB, the section of model output labeled "Time Averaged Volume Concentration: Concentration Contour Parameters" summarizes the model output that reflects the (desired) effect of plume meander (but does not include effects such as integrating the concentration time history). Using Equations in the SLAB User's Manual (and also reproduced in the model output), the necessary input to HABIT is calculated. (The SLAB output is written so that all parameters are a function of the reported time, and the concentration time history is calculated using $x = X_p$, $y = 0$, and $z = Z_p$.)

7. Intake Height

One of the parameters set by the user in HABIT is the intake height. Users can choose to set the air intake height at ground level to provide a conservative estimate of the contaminant concentration that could be entering the control room. If the vertical extent of the contaminant cloud is significantly higher than the actual intake height, this conservative assumption could be unimportant, but when this is not the case, the contaminant concentration estimates could be substantially different.

To illustrate the impact of such an assumption, consider the predicted impact of the sudden rupture of a 1 ton (907 kg) chlorine cylinder. Initially, the liquid chlorine is at ambient temperature, but after release, the chlorine is expected to form an aerosol cloud, and following the approach outlined above, the aerosol cloud would have an initial density of 20.1 kg/m^3. If the air intake is located 1.5 km from the release point and is 15 m above grade, the maximum ground level concentration is predicted by DEGADIS to be 108 ppm, but the maximum concentration at the air intake would only be 20 ppm.

For denser-than-air gases (and stable atmospheric conditions), there is justification for recognizing that using the concentration at ground level is actually inappropriate to estimate the concentration at an elevated location. As depicted in Figure 1, the phenomenon of a dividing streamline for a contaminant plume flow around an obstacle has been observed in nature and studied in wind tunnels (e.g., Snyder et al., 1985). The dividing streamline is the horizontal surface on a hill or obstacle below which the plume will travel laterally around the obstacle (orange arrow), and above which the plume will travel over the hill or obstacle (green arrow). For this application, the contaminant concentration at the elevation of the dividing streamline would be the maximum concentration that could be predicted to occur at the intake height.

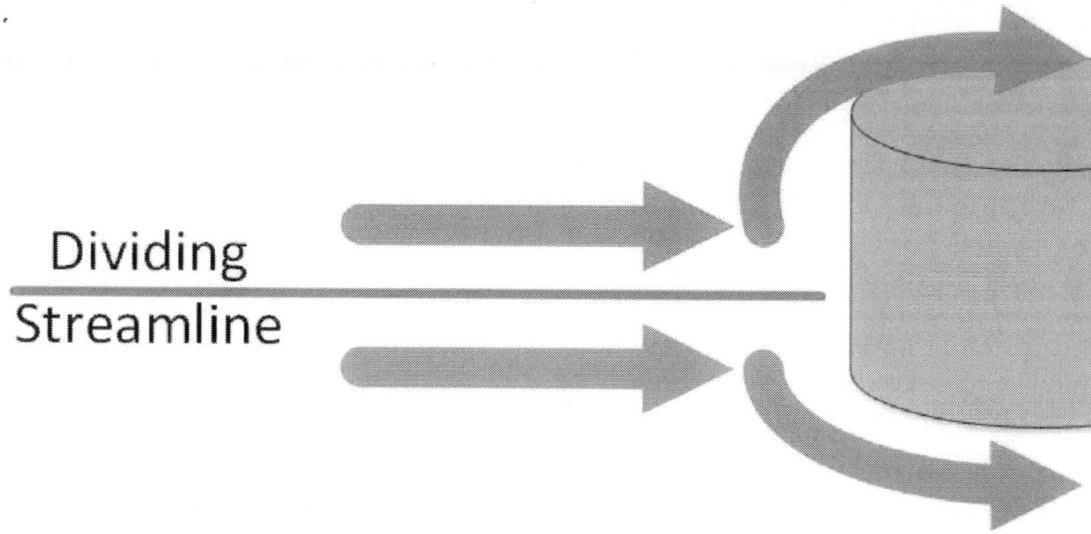

Figure 1. Below the Dividing Streamline, flow is around the obstacle, but above, flow is over the obstacle.

For passive gas releases, the height of the dividing streamline (H_s) is described using Sheppard's formula as:

$$\frac{1}{2}\rho_a U_a(H_s)^2 = g \int_{H_s}^{H} (H-z)\left(-\frac{d\rho}{dz}\right) dz \tag{3}$$

where $U_a(H_s)$ is the ambient wind speed evaluated at H_s. Snyder et al.(1985) show that Sheppard's equation is consistent with wind tunnel studies. Note that Equation (1) implies that $H_s = 0$ for the convective boundary layer (daytime conditions) if the release is passive. For stable boundary layer conditions, $H_s \geq 0$. Dispersion models that account for denser-than-air gas effects can be used to estimate ($-d\rho/dz$).when the contaminant cloud is present. For the example discussed above, H_s was 19.3 m, so using the ground level concentration seems unjustified. However, when the dividing streamline height is significantly less than the air intake height, the calculated concentration at the air intake may be too low because of higher concentrations at the dividing streamline may be lofted to the air intake as predicted by Equation (1).

8. Conclusion

DEGADIS and SLAB have been incorporated into HABIT for the purpose of considering the effects of denser-than-air contaminant cloud dispersion. The Britter and McQuaid criteria for predicting the importance of denser-than-air effects on dispersion have been implemented as a condition for using SLAB and DEGADIS.

For two-phase flashing releases, some liquid rainout may occur, but this case can be treated conservatively by assuming the liquid rainout is unimportant. This release scenario is simulated in SLAB as if the flashing release occurred from a ground level, area source as a vertical jet. For this scenario, SLAB and DEGADIS use the area of the release to predict several source parameters assuming the vertical jet velocity is the same as the ambient wind speed. As modeled by EXTRAN, liquid leaks and tank bursts that form an evaporating pool can also include a fraction of the release that flashes to vapor, and this (rapid) vapor release has been accounted for in the DEGADIS simulations. For pressurized gas releases, the gas temperature after depressurization is reduced during the process of adiabatic expansion against atmospheric pressure.

For denser-than-air gases (and stable atmospheric conditions), there is justification for recognizing that using the concentration at ground level is actually inappropriate to estimate the concentration at an elevated location. The dividing streamline is the horizontal surface on a hill or obstacle below which the plume will travel laterally around the plume, and above which the plume will travel over the hill or obstacle. For considering the concentration at the air intake, the contaminant concentration at the elevation of the dividing streamline would be the maximum concentration that could be predicted to occur at the intake height. For a passive release in convective boundary layer conditions, the ground level concentration may be appropriate for use at the air intake concentration, but for stable boundary layer conditions or when a denser-than-air cloud is present, this would not be true. Conversely, the predicted concentration at the air intake could underestimate the concentration because of the effect of the dividing streamline for stable boundary layers or when a denser-than-air contaminant cloud is present. Additional study of this aspect of the problem seems appropriate.

References

Britter, R., and J. McQuaid, Workbook on the dispersion of Dense Gases, Health and Safety Executive Report 17/1988, Sheffield, UK, 1988.

Ermak, D.L., User's Manual for SLAB: An Atmospheric Dispersion Model for Denser-than-Air Releases, UCRL-MA-105607, Lawrence Livermore National Laboratory, 1990.

Ramsdell, J.V., EXTRAN: A Computer Code for Estimating Concentrations of Toxic Substances at Control Room Intakes, NUREG/CR-5656, PNL-7510, U.S. Nuclear Regulatory Commission, 1991.

Snyder, W.H., R.S. Thompson, R.E. Eskridge, R.E. Lawson, I.P. Castro, and J.T. Lee, The structure of strongly stratified flow over hills: dividing-streamline concept," *Journal of Fluid Mechanics*, v. 152, 249-288, 1985.

Spicer, T., and J. Havens, User's Guide for the DEGADIS 2.1 Dense Gas Dispersion Model, EPA-450/4-89-019, U.S. Environmental Protection Agency Office of Air Quality Planning and Standards, 1989.

Witlox, H. W. M., M. Harper, A. Oke, P. Bowen, and P. Kay, "Sub-cooled and flashing liquid jets and droplet dispersion I. Overview and model implementation/validation," *Journal of Loss Prevention in the Process Industries, 23*, 831-842, 2010.

MODELING OF SELF-HEATING IN ANION EXCHANGE COLUMNS FOR PLUTONIUM RECOVERY

Dr. James E. Laurinat
Savannah River National Laboratory
Aiken, South Carolina

Abstract

A finite element code has been developed to model radiolytic self-heating in anion exchange columns. Anion exchange columns are used to extract plutonium from nitric acid solutions as part of the Purex process. When columns are used to extract Pu-238, significant rates of radiolytic heating pose a risk that column temperatures could rise to the point where a runaway resin degradation reaction occurs. A runaway degradation reaction can lead to a potentially violent thermal excursion, where gas generation from the degradation reaction causes the column to eruct. The finite element code focuses on predicting the maximum temperature due to radiolytic heating in order to ensure that this maximum temperature remains below the temperature required to initiate a reaction excursion. The code also can be used to calculate the cumulative radiological dose to the resin.

The finite element code calculates concentration profiles and temperatures for normal operation of a column, including loading, washing, and elution of the resin, and for natural circulation during flow stoppages. Resin loading is modeled by combining a chemical equilibrium absorption relation with models for axial dispersion and solid phase diffusion. Desorption is modeled by combining the axial dispersion model with a shrinking core diffusion model. Natural convection calculations are based on the Ergun relation for flow in a packed bed, using pressure gradients obtained from temperature and composition differences in the column solution.

The model was applied to calculate maximum temperatures in anion exchange columns used to extract Pu-238 for the Cassini mission at the Savannah River Site (SRS) and for a Pu-238 laboratory-scale column at Los Alamos National Laboratory (LANL). More recently, the model was used to predict doses to the resin for feeds containing plutonium, americium, and curium. For this calculation, it was assumed that the resin absorbs plutonium much more strongly than either americium or curium. Results are presented for all three studies.

Nomenclature

C_{Pu}	total plutonium concentration adsorbed on the resin, in resin pores, and in the bulk solution, kg/m^3
c	fractional concentration of elution acid (250 mol/m^3), dimensionless
$c_{p,f}$	solution heat capacity, J/kg/K
$c_{p,bed}$	bulk heat capacity for the resin bed, J/kg/K
D_i	distribution coefficient for ith element with respect to Pu
D_{mix}	dispersion coefficient for mixing due to density gradients, m^2/s
D_{part}	particle diffusivity for diffusion of Pu in the resin bead, m^2/s

D_{pore}	bulk pore diffusivity for diffusion of Pu in the resin bead, normalized to account for differences between solid and solution concentrations, m^2/s
D'_{pore}	bulk pore diffusivity for diffusion of Pu in the resin bead, m^2/s
d	column diameter, m
d_p	average resin bead diameter, m
f	bulk concentration of plutonium adsorbed onto the resin, kg/m^3
f_{feed}	bulk concentration of plutonium adsorbed onto the resin when the resin is fully loaded, kg/m^3
f_0	bulk concentration of plutonium adsorbed onto the resin at the end of the decon wash, kg/m^3
G	during loading, bulk rate of transfer of plutonium from bulk solution to the resin, $kg/m^3/s$
G_a	during elution, bulk rate of transfer of plutonium from the resin to the bulk solution due to desorption, $kg/m^3/s$
G_b	during elution, bulk rate of diffusion of plutonium from the bulk solution to the resin pore solution, $kg/m^3/s$
g	gravitational acceleration, m/s^2
K_1	equilibrium constant for the solubility of plutonium from nitric acid solutions on Ionac 641 resin, (mol/m^3)
k_{bed}	bulk thermal conductivity for the resin bed, J/m/s/K
Q_{decay}	mass rate of heat generation due to radioactive decay of plutonium isotopes, W/kg
R	column radius, m
R_p	average resin bead radius, m
$R_{p,1}$	during elution, radius of the undesorbed core of resin bead, m
r	radial distance from the center of the column, m
r_p	radial distance from the center of the resin bead, m
T	bulk temperature, K
t	time for heat transfer and mass transfer to bulk solution, s
t'	characteristic time for adsorption with continuous contact between the loading solution and the resin bead surfaces, s
t_p	time for adsorption within a resin bead, s
$\langle t_p \rangle$	average residence time for adsorption at a given axial level, s
$t_{p,a}$	time for desorption within a resin bead, s
$\langle t_{p,a} \rangle$	average residence time for desorption at a given axial level, s
$t_{p,b}$	time for interchange between the bulk solution and resin bead pores, s
$\langle t_{p,b} \rangle$	average residence time for interchange between the bulk solution and resin bead pores at a given axial level, s
v_c	characteristic velocity for axial mixing due to density gradients, m/s

$v_{c,eff}$	effective velocity for density gradient mixing, m/s
v_r	superficial radial velocity in the column, m/s
v_z	superficial axial velocity going from the top to the bottom of the column, m/s
x	during loading and the decon wash, bulk concentration of plutonium in the bulk solution, kg/m^3
x_a	during desorption, local concentration of plutonium in the resin bead pores, kg/m^3
x_{feed}	concentration of plutonium in the feed solution, kg/m^3
x_{max}	maximum plutonium concentration in the bulk solution, kg/m^3
x_{sat}	plutonium concentration in the bulk solution at saturation, kg/m^3
y	during elution, bulk concentration of plutonium in the resin bead pores due to diffusion from the bulk solution, kg/m^3
$\bar{y}$	during desorption, concentration of plutonium in the pores at the resin bead surface, kg/m^3
z	axial distance from the top of the resin bed, m
z_0	axial location at which the dispersion coefficient for density gradient mixing is evaluated, m
α_r	radial dispersion coefficient, m^3/s
α_z	axial dispersion coefficient, m^3/s
ε_i	volume fraction of column occupied by liquid in interparticle space
$\varepsilon_{i,1}$	fraction of the total resin bed volume occupied by the resin bead pores before desorption of plutonium, dimensionless
$\varepsilon_{i,2}$	fraction of the total resin bed volume occupied by the resin bead pores after desorption of plutonium, dimensionless
$\varepsilon_{i,2,0}$	fraction of the total resin bed volume occupied by the resin bead pores after all plutonium has desorbed, dimensionless
ε_o	fraction of the total resin bed volume occupied by the bulk solution, dimensionless
η	factor to account for the difference between the pseudo-steady-state shrinking core approximation and the exact shrinking core model solution for the total desorption time, dimensionless
μ	solution dynamic viscosity, kg/m/s
ρ_f	solution density, kg/m^3
$\bar{\rho}_f$	average solution density at a given axial level, kg/m^3
ρ_{bed}	bulk density of the resin bed, kg/m^3
τ_{part}	time for a resin bead to become fully loaded with plutonium due to particle diffusion, s
τ_{pore}	time for complete desorption of a resin bead by pore diffusion, s

Introduction

The original impetus to develop a model for the recovery of plutonium in anion exchange columns was to assess the change in column temperatures for increased plutonium loading on the column. Historically, at SRS, an anion exchange column was used to recover and recycle Pu-238 from scrap solutions. When recovery of Pu-238 to make product for the Cassini mission was started, an increase in column loading was desired. (The Cassini mission was a joint endeavor of the National Aeronautics and Space Administration (NASA), the European Space Agency (ESA), and the Italian space agency Agenzia Spaziale Italiana (ASI), to explore Saturn and its moons. Pu-238 was used in radioisotope thermoelectric generators (RTGs) to power the Cassini satellite.)

The Pu-238 column load was limited to prevent thermal degradation and disintegration of the resin bed. At low temperatures, resin bed temperatures increase primarily due to decay heat from Pu-238. At temperatures exceeding about 388 K, an exothermic reaction between nitric acid and resin predominates. This reaction causes a potentially violent temperature excursion. Operating limits stipulated a maximum column temperature of 333 K to protect against the exothermic nitric acid-resin reaction.

The greatest potential for a resin bed temperature excursion occurs during flow interruptions. If the flow is stopped during loading, washing, or elution, the temperature rises due to heat from the decay of Pu-238 until conditions for a temperature excursion are met. The temperature rise is nearly adiabatic; calculations showed that natural convection to the ambient air is not effective in cooling the resin bed.

The original model evaluated the effects of increases in column loading on column temperatures during flow interruptions, as a function of time, for a uniform velocity flow through the column.[1] Subsequently, at the request of Los Alamos National Laboratory, the model was extended to include natural convection in order to model temperature increases inside a column during a flow stoppage.[2,3,4] Finally and more recently, the original, one-dimensional flow model was coupled with a dose calculation to estimate the total dose to the resin for a feed containing plutonium, americium, and curium, along with cesium and strontium fission products.

The following sections describe the mass transfer model, which was developed for one-dimensional uniform flow, and the heat transfer aspects of the natural convection model for the resin bed, which is applied assuming two-dimensional axisymmetric flow. Results from both the mass and heat transfer calculations are reported. The mass transfer and heat transfer model descriptions are followed by a brief overview of the resin dose calculation.

Mass Transfer Model

<u>Ion Exchange Chemistry</u>

In the ion exchange column, plutonium is selectively adsorbed on an anionic resin from an 8000 mol/m^3 HNO_3 solution and is eluted with 250 mol/m^3 HNO_3. Historically, plutonium ion exchange columns used Dowex-1™ X-4 ion exchange resin. At the time the original model was developed, the column used Ionac 641™ macroporous resin converted to a nitrate form. Later, both the LANL laboratory-scale column and the SRS mixed actinide ion exchange column

™ Dowex-1 is a registered trademark of DowDuPont of Midland, Michigan.

™ Ionac 641 is a registered trademark of LANXESS Sybron Chemicals, Inc., of Birmingham, New Jersey.

used Reillex HPQ™ pyridine-based resin. For all three resins, the following reaction occurs on the resin surfaces,[5] which are assumed to be evenly distributed throughout the resin beads.

$$R_2Pu(NO_3)_6\,(solid) \Leftrightarrow PuNO_3^{3+}(aq) + 3NO_3^-(aq) + 2RNO_3(solid)$$

where R denotes a chemically active site on the resin substrate.

Equilibrium solution concentrations were calculated by combining the solubility product $K_1 = \left[PuNO_3^{3+}\right]\left[NO_3^-\right]^3$ with the equilibrium product for the aqueous phase complexation reaction[5]

$$Pu^{4+}(aq) + NO_3^-(aq) \Leftrightarrow PUNO_3^{3+}(aq)$$

The solubility product depends on the type of resin. Figure 1 depicts the solubility as a function of nitric acid concentration for both Dowex-1™ and Ionac 641™.

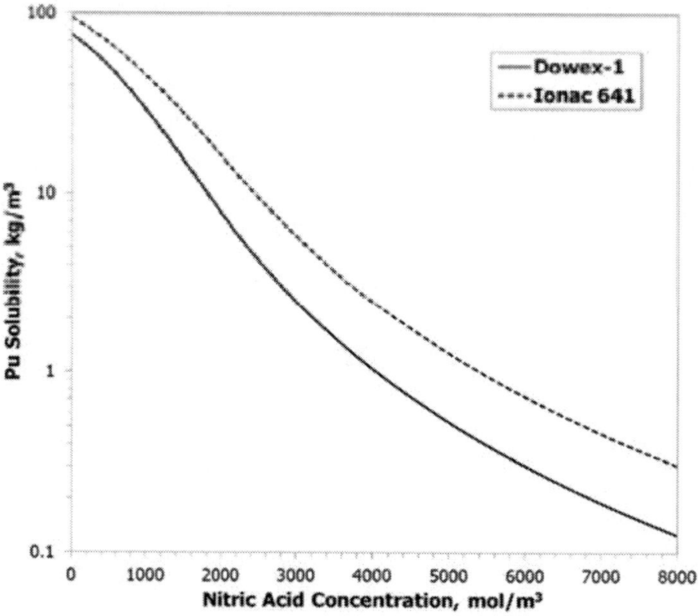

Figure 1. Saturated Equilibria for Dowex-1™ and Ionac 641™ Resins in Nitric Acid

Mass Transfer Model Development

Separate mass transfer models were developed for loading and washing and for elution. Both models use one-dimensional, transient mass conservation equations, with explicit terms to account for adsorption or elution of plutonium into or out of the resin beads. These models divide the total plutonium concentration, C_{Pu}, into a bulk flow concentration, x, a resin bead pore solution concentration, y, and an adsorbed concentration, f:

$$C_{Pu} = x + y + f \tag{1}$$

All concentrations are based on the bulk volume of the resin bed.

™ Reillex HPQ is a registered trademark of Vertellus Specialty Chemicals, Inc., of Indianapolis, Indiana.

Adsorption Mass Transfer

The mass balances for adsorption and washing take the form

$$\frac{\partial x}{\partial t} = -v_z \frac{\partial x}{\partial z} + \alpha_z \frac{\partial^2 x}{\partial z^2} \qquad (2)$$

and

$$\frac{\partial f}{\partial t} = G \qquad (3)$$

The axial dispersion coefficient is calculated using a mixing length of two bead diameters.[6] This gives

$$\alpha_z = 2 d_p v_z \qquad (4)$$

Diffusion to the resin bead surfaces controls the rates of adsorption and elution. Because the plutonium ionic species react rapidly at the resin surface, adsorption takes place at the outer surfaces of the resin beads before the plutonium can penetrate to the interior of the beads. Consequently, solid-phase, or particle, diffusion controls the rate of adsorption.[5] The particle diffusion model is analogous to a molecular diffusion model, with a particle diffusivity and the bulk adsorbed concentration replacing the ordinary diffusivity and the solution concentration.

An approximate solution for diffusion through a spherical particle is[7]

$$\frac{t_p}{\tau_{part}} = -\ln\left(1 - \left(\frac{f}{f_{feed}}\right)^2\right) \qquad (5)$$

where

$$\tau_{part} = \frac{d_p^{\,2}}{4\pi^2 D_{part}} \qquad (6)$$

The variable t_p represents the time needed for a single resin bead to adsorb a given amount of plutonium. The collective adsorption time for resin beads at a particular bed level, $\langle t_p \rangle$, is an average of these individual adsorption times, weighted by the adsorbed fraction f. The integral representation of this weighted average is:

$$\langle t_p \rangle = \frac{f_{feed}}{f} \int_0^{t_p} \frac{f}{f_{feed}} dt_p \qquad (7)$$

The time required for the loading calculation is the average time it takes for a level of the resin bed to adsorb plutonium. The calculation of this loading time is based on the stipulation that the loading profile is fully developed. If the resin beads are in continuous contact with the loading solution, then, for a fully-developed profile, the average adsorption time $\langle t_p \rangle$ and the loading time t' are related by the characteristic equation,

$$\langle t_p \rangle \Delta f = f \Delta t' \qquad (8)$$

At the loading front, however, only a portion of the bulk solution is in saturated equilibrium with the adsorbed plutonium on the resin surfaces. The mixing of saturated and unsaturated solution occurs at a macroscopic rather than a molecular level, because axial dispersion predominates over molecular diffusivity. Consequently, this mixing can be modeled by assuming that the bulk flow is segregated into one fraction that contains dissolved plutonium and another that does not. Under this assumption, the actual loading time, t, and the loading time for continuous contact, t', are related by

$$\Delta t' = \frac{x}{x_{feed}} \Delta t \tag{9}$$

The preceding equations combine to yield the following expression for the loading mass transfer rate

$$G = f_{feed} \frac{x}{x_{feed}} \frac{\left(\frac{f}{f_{feed}}\right)^2}{\ln\left(\frac{f_{feed}+f}{f_{feed}-f}\right) - 2\frac{f}{f_{feed}}} \frac{1}{\tau_{part}} \tag{10}$$

As $f \rightarrow 0$,

$$G = f_{feed} \frac{x}{x_{feed}} \frac{1.5 f_{feed}}{f} \frac{1}{\tau_{part}} \tag{11}$$

Elution Mass Transfer

During elution, dilute acid diffuses inward from the resin bead surface to displace more concentrated acid, and plutonium nitrate species diffuse outward toward the surface. Thus, the acid concentration increases and the plutonium nitrate concentration decreases going toward the center of the bead. This implies that a saturated plutonium nitrate solution exists at only one radial location within the bead, provided the adsorbed plutonium concentration is sufficient to locally saturate the pore solution. Plutonium nitrate species diffuse outward from this surface.

The mass balances for elution are:

$$\frac{\partial c}{\partial t} = \frac{\varepsilon_o}{\varepsilon_o + \varepsilon_{i,1}} \left(-v_z \frac{\partial c}{\partial z} + \alpha_z \frac{\partial^2 c}{\partial z^2} \right) \tag{12}$$

$$\frac{\partial x}{\partial t} = -v_z \frac{\partial x}{\partial z} + \alpha_z \frac{\partial^2 x}{\partial z^2} + G_a - G_b \tag{13}$$

$$\frac{\partial f}{\partial t} = -G_a \tag{14}$$

and

$$\frac{\partial y}{\partial t} = G_b \tag{15}$$

where c is the fraction of the bulk solution that is 250 mol/m^3 elution acid, G_a is the rate of plutonium desorption, and G_b is the rate of diffusion of plutonium from the bulk solution into desorbed resin beads. The factor $\frac{\varepsilon_o}{\varepsilon_o + \varepsilon_{i,1}}$, which is the ratio of the bulk solution volume to the total solution volume before desorption, accounts for interchange of elution acid between the bulk and pore solutions. Use of this factor is based on a quasi-equilibrium approximation. In other words, diffusion of nitric acid is assumed to be rapid compared with diffusion of plutonium nitrate species, so that the pore and bulk solution nitric acid concentrations are equal while plutonium diffuses.

Separate rate models are used for desorption and interchange of plutonium between the bulk and pore solutions, because the desorption model transfers plutonium directly from the adsorbed solid to the bulk solution, without allowing for any plutonium in the pore volume. A shrinking core approximation describes desorption of plutonium from the resin beads.[5]

If the pore diffusivity is assumed to be constant, the pore diffusion equation becomes

$$\frac{\partial x_a}{\partial t_{p,a}} = \frac{D'_{pore}}{r_p^2}\frac{\partial}{\partial r_p}\left(r_p^2 \frac{\partial x_a}{\partial r_p}\right) \tag{16}$$

A pseudo-steady approximation is applied to eliminate time dependent terms:

$$\frac{d}{dr_p}\left(r_p^2 \frac{dx_a}{dr_p}\right) = 0 \tag{17}$$

The steady-state approximation is valid, provided that the concentration of plutonium adsorbed on the surface is greater than the bulk concentration in a solution in saturated equilibrium with the adsorbed plutonium. Errors introduced by this approximation are accounted for later in this derivation.

Equation 16 is solved with the boundary conditions

$$x_a = x_{max} \text{ at } r_p = R_{p,1} \tag{18}$$

$$\text{and } x_a = 0 \text{ at } r_p = R_p \tag{19}$$

The pseudo-state-state solution is

$$x_a = x_{max}\frac{R_{p,1}\left(R_p - r_p\right)}{r_p\left(R_p - R_{p,1}\right)} \tag{20}$$

This solution is useful only when expressed in terms of a fractional desorption. To convert this result to a useful form, a mass balance is applied to equate the shrinkage of the undissolved core region with the diffusion rate at the bead surface. This takes the form

$$4\pi R_p^2 D'_{pore}\left.\frac{dx_a}{dr_p}\right|_{r_p=R_p} = f_0 \frac{d}{dt_p}\left(\frac{4}{3}\pi R_{p,1}^3\right) \tag{21}$$

The initial condition for this equation is

$$R_{p,1} = R_p \text{ when } t_{p,a} = 0, \tag{22}$$

and the final condition is

$$R_{p,1} = 0 \text{ when } t_{p,a} = \tau_{pore} \tag{23}$$

The fractional desorption is defined by

$$\frac{f}{f_0} = \left(\frac{R_{p,1}}{R_p}\right)^3 \tag{24}$$

Solution of the preceding equations yields the following expression for the rate of desorption during elution:

$$\frac{t_{p,a}}{\tau_{pore}} = 1 + 2\frac{f}{f_0} - 3\left(\frac{f}{f_0}\right)^{\frac{2}{3}} \tag{25}$$

where

$$\tau_{pore} = \frac{d_p^{\,2} f_0}{24 D'_{pore} x_{max}} \tag{26}$$

A modified pore diffusivity is defined to incorporate differences between the adsorbed solid and saturated solution concentrations of plutonium:

$$D_{pore} = \frac{D'_{pore} x_{max}}{f_0} \tag{27}$$

In terms of this modified diffusivity,

$$\tau_{pore} = \frac{d_p^{\,2}}{24 \eta D_{pore}} \tag{28}$$

An effectiveness factor η of 0.75 was added to this equation to account for errors introduced by the pseudo-state-state approximation.

As in the adsorption rate analysis, the collective adsorption time for resin beads at a particular bed level, $\langle t_{p,a} \rangle$, is calculated as a weighted average of individual adsorption times, $t_{p,a}$, from the integral,

$$\langle t_{p,a} \rangle = \frac{f_0}{f} \int_0^{t_{p,a}} \frac{f}{f_0} dt_{p,a} \tag{29}$$

This integral gives

$$\frac{\langle t_{p,a} \rangle}{\tau_{pore}} = \frac{f}{f_0} - \frac{6}{5}\left(\frac{f}{f_0}\right)^{\frac{2}{3}} + \frac{f_0}{5f} \tag{30}$$

The expression for the rate of desorption is obtained by differentiating Equation 30. This gives

$$\frac{1}{f_0} \frac{df}{d\langle t_{p,a} \rangle} = \frac{1}{\tau_{pore}\left(1 - \frac{4}{5}\sqrt[3]{\frac{f_0}{f}} - \frac{1}{5}\left(\frac{f_0}{f}\right)^2\right)} \tag{31}$$

Equation 31 gives the collective rate of desorption for resin beads continuously in contact with a bulk solution of 250 mol/m^3 elution acid that does not contain plutonium. Correction factors must be added to account for variations in acid and plutonium concentrations in the bulk solution. To derive a correction factor for acid concentration, it is assumed that during desorption the bulk flow is segregated into one fraction, c, that contains dilute elution acid (250 mol/m^3) and another, $1-c$, that contains wash acid (8000 mol/m^3). This implies that changes in the collective desorption time and the elapsed time are related by

$$\Delta\langle t_{p,a} \rangle = c\Delta t \tag{32}$$

To correct Equation 31 for the plutonium concentration gradient, the average plutonium concentration in the pores at the resin bead surface, y, must be calculated. This concentration is obtained from a quasi-equilibrium relation, in which the rate of diffusion of plutonium to the bulk solution varies about a mean of zero. To derive this relation, it is assumed that the elution acid is poorly mixed in the bulk solution, so that it segregates into a fraction $\frac{x}{x_{max}}$ that is in saturated equilibrium with the adsorbed plutonium and a fraction $c-\frac{x}{x_{max}}$ that contains no plutonium. It follows that, if the diffusion resistance of the bulk solution is negligible, plutonium diffuses into the resin bead from a saturated bulk solution a fraction of time $\frac{x}{x_{max}}$ and diffuses from the resin bead into elution acid that contains no plutonium a fraction of time $c-\frac{x}{x_{max}}$. The quasi-equilibrium relation is obtained by integrating the transient diffusion equation. Such an integration shows that the mass transfer is proportional to the concentration difference and the square root of the contact time. In this instance, then,

$$\left(x_{max}-\bar{y}\right)\sqrt{\frac{x}{x_{max}}}-\bar{y}\sqrt{c-\frac{x}{x_{max}}}=0 \tag{33}$$

Solution of this equation gives the concentration gradient correction factor,

$$1-\frac{\bar{y}}{x_{max}}=\frac{\sqrt{c-\frac{x}{x_{max}}}}{\sqrt{\frac{x}{x_{max}}}+\sqrt{c-\frac{x}{x_{max}}}} \tag{34}$$

The mass transfer rate for desorption, including the correction factors for the acid concentration and the plutonium concentration gradient, is

$$G_a=-\left(1-\frac{\bar{y}}{x_{max}}\right)\frac{df}{d\langle t_{p,a}\rangle}\frac{d\langle t_{p,a}\rangle}{dt} \tag{35}$$

Evaluation of Equation 35 yields the rate expression,

$$G_a=\frac{-cf_0\left(\frac{\sqrt{c-\frac{x}{x_{max}}}}{\sqrt{\frac{x}{x_{max}}}+\sqrt{c-\frac{x}{x_{max}}}}\right)}{\tau_{pore}\left(1-\frac{4}{5}\sqrt[3]{\frac{f_0}{f}}-\frac{1}{5}\left(\frac{f_0}{f}\right)^2\right)} \tag{36}$$

The desorption model does not account for diffusion of plutonium into and out of the resin bead pores as the elution concentration spike travels down the column. The diffusion equation for this interchange is based on the approximate solution for particle diffusion in a sphere,[5] which uses a quadratic concentration gradient. Similarly, pore diffusion between the bulk and pore solutions is modeled with quadratic gradients. Based on the assumption that the bulk solution is poorly mixed at the molecular level, it is assumed that the bulk solution is segregated into one volume fraction in which the plutonium concentration is in saturated equilibrium with the adsorbed plutonium and another that does not contain any plutonium. The net diffusion for such a segregated mixture can be expressed by

$$\frac{1}{\varepsilon_{i,2}^{2}}\frac{dy^2}{dt_{p,b}}=\frac{1-\varepsilon_0}{\varepsilon_{i,2}}\frac{4\pi^2 D_{pore}}{d_p^{2}}\left(\frac{x}{x_{max}}\left(\left(\frac{x_{max}}{\varepsilon_0}\right)^2-\left(\frac{y}{\varepsilon_{i,2}}\right)^2\right)-\left(1-\frac{x}{x_{max}}\right)\left(\frac{y}{\varepsilon_{i,2}}\right)^2\right) \tag{37}$$

This simplifies to

$$\frac{1}{\varepsilon_{i,2}^{2}}\frac{dy^2}{dt_{p,b}}=\frac{1-\varepsilon_o}{\varepsilon_{i,2}}\frac{4\pi^2 D_{pore}}{d_p^{2}}\left(\frac{x_{max}x}{\varepsilon_o^{2}}-\frac{y^2}{\varepsilon_{i,2}^{2}}\right) \tag{38}$$

where the pore liquid concentration is based on the volume fraction in the pores of the eluted portion of the resin beads, $\varepsilon_{i,2}$ The factor $\frac{1-\varepsilon_o}{\varepsilon_{i,2}}$ appears because the pore plutonium concentration is normalized with respect to the pore volume, whereas the pore diffusivity applies to the entire resin bead volume.

The eluted pore volume fraction is related to the pore volume in a completely eluted bead, $\varepsilon_{i,2,0}$, by

$$\varepsilon_{i,2}=\left(1-\frac{f}{f_0}\right)\varepsilon_{i,2,0} \tag{39}$$

The initial loading has been used in place of the full equilibrium loading in this calculation, because it gives the correct initial condition: i.e., the eluted pore volume fraction must be zero at the beginning of the elution.

Time variations of the eluted pore volume fraction have been neglected in Equation 38. This simplification does not introduce significant errors, since the rate of desorption is slower than the rate of interdiffusion of plutonium between the pores and the bulk liquid.

The collective adsorption time for resin beads at a particular bed level, $\langle t_{p,b}\rangle$, is calculated as a weighted average of individual adsorption times, $t_{p,b}$,from the integral,

$$\langle t_{p,b}\rangle=\frac{f_0}{f}\int_0^{t_{p,b}}\frac{f}{f_0}dt_{p,b} \tag{40}$$

Solution of this integral gives

$$\langle t_{p,b}\rangle=\frac{d_p^{2}}{4\pi^2 D_{pore}\varepsilon_{i,2}(1-\varepsilon_o)}\left(\frac{\varepsilon_{i,2}\sqrt{x_{max}x}}{\varepsilon_o y}\ln\left(\frac{\frac{\sqrt{x_{max}x}}{\varepsilon_o}+\frac{y}{\varepsilon_{i,2}}}{\frac{\sqrt{x_{max}x}}{\varepsilon_o}-\frac{y}{\varepsilon_{i,2}}}\right)-2\right) \tag{41}$$

The expression for the rate of desorption is obtained by differentiating Equation 41 to get

$$\frac{dy}{d\langle t_b\rangle}=\frac{4\pi^2 D_{pore}(1-\varepsilon_o)\left(\frac{y^2}{\varepsilon_{i,2}^{2}}\right)\left(\frac{x_{max}x}{\varepsilon_o^{2}}-\frac{y^2}{\varepsilon_{i,2}^{2}}\right)}{d_p^{2}\left(2\frac{x_{max}xy}{\varepsilon_o^{2}\varepsilon_{i,2}}-\frac{\sqrt{x_{max}x}}{\varepsilon_o}\left(\frac{x_{max}x}{\varepsilon_o^{2}}-\frac{y^2}{\varepsilon_{i,2}^{2}}\right)\ln\left(\frac{\frac{\sqrt{x_{max}x}}{\varepsilon_o}+\frac{y}{\varepsilon_{i,2}}}{\frac{\sqrt{x_{max}x}}{\varepsilon_o}-\frac{y}{\varepsilon_{i,2}}}\right)\right)} \tag{42}$$

The mass transfer rate is expressed in terms of the actual elapsed time, so that

$$G_b = \frac{dy}{d\langle t_{p,b}\rangle}\frac{d\langle t_{p,b}\rangle}{dt} \quad (43)$$

Changes in the time for diffusion interchange equal changes in the actual elapsed time, $\Delta\langle t_{p,b}\rangle = \Delta t$, so that

$$G_b = \frac{4\pi^2 D_{pore}(1-\varepsilon_o)\left(\frac{y^2}{\varepsilon_{i,2}^{\,2}}\right)\left(\frac{x_{max}x}{\varepsilon_o^{\,2}} - \frac{y^2}{\varepsilon_{i,2}^{\,2}}\right)}{d_p^{\,2}\left(2\frac{x_{max}xy}{\varepsilon_o^{\,2}\varepsilon_{i,2}} - \frac{\sqrt{x_{max}x}}{\varepsilon_o}\left(\frac{x_{max}x}{\varepsilon_o^{\,2}} - \frac{y^2}{\varepsilon_{i,2}^{\,2}}\right)ln\left(\frac{\frac{\sqrt{x_{max}x}}{\varepsilon_o}+\frac{y}{\varepsilon_{i,2}}}{\frac{\sqrt{x_{max}x}}{\varepsilon_o}-\frac{y}{\varepsilon_{i,2}}}\right)\right)} \quad (44)$$

As $\frac{y}{\varepsilon_{i,2}} \to \frac{\sqrt{x_{max}x}}{\varepsilon_o}$, this reduces to

$$G_b = \frac{4\pi^2 D_{pore}(1-\varepsilon_o)}{d_p^{\,2}}\left(\frac{\sqrt{x_{max}x}}{\varepsilon_o} - \frac{y}{\varepsilon_{i,2}}\right) \quad (45)$$

Comparison of Mass Transfer Model Predictions with Measurements

Model predictions were compared with a previously measured column loading profile and with results from two laboratory-scale test elutions. In the first elution, a short 0.092-m column was nearly fully loaded with plutonium. Results from this elution and the measured profile were used to determine the saturated equilibrium plutonium concentration, the maximum plutonium loading, and the particle diffusivity. The second elution, in which a longer 0.375-m column was partially loaded, approximated conditions in the production column. Results from both elutions were used to estimate the pore diffusivity and volume fractions for the bulk and pore solutions.

The loading, washing, and elution times for the second test elution were shortened so that this elution could be completed in one working day. Although this test was abbreviated, the elution time was sufficient to allow removal of most of the plutonium from the column. During typical column elutions, the head cut takes 1.2 bed volumes, the heart cut takes 1.5 bed volumes, and the tail cut takes 1.9 bed volumes. The total elution volume for the second test elution was approximately equal to the head plus heart cuts for typical column operation.

Figure 2 compares measured and predicted loading profiles. A particle diffusivity of 2E-13 m^2/s was used to fit the reported data. This diffusivity is the same magnitude as that measured for Dowex™ X-8 resin.[5]

From the results of the first test elution, it was determined that the maximum loading at typical column operating conditions is 45 kg/m^3. As Figure 3 shows, the model accurately predicts the cumulative plutonium discharge with this maximum loading concentration and the particle diffusivity determined from the loading profile analysis.

Figures 4 and 5 compare model predictions with measured rates of discharge of plutonium during both test elutions. A pore diffusivity of 1.0E-10 m^2/s and pore volume fractions of 0.132 prior to desorption of plutonium from the resin and 0.160 after desorption were used to fit the model to the test data. The pore diffusivity is about 100 times greater than reported values for anionic resins.[5] The pore volume fractions correspond to resin bead porosities of 0.198 before desorption and 0.240 after desorption. The porosity increases during desorption due to swelling of the resin and loss of Pu•4NO_3 by the reaction[5]

$$R_2Pu(NO_3)_6\,(solid) \Leftrightarrow PuNO_3^{3+} + 3NO_3^- + 2RNO_3(solid)$$

The solubility for Ionac 641™ resin was estimated from eluate plutonium concentrations measured during the first test elution. Before the plutonium concentration spike discharged from the column, the plutonium concentration asymptotically approached a minimum of 0.0106 g/L at a nitric acid concentration of 7850 mol/m^3. The plutonium concentration then rapidly peaked at an apparently constant concentration of 26.3 kg/m^3 at a nominal elution acid concentration of 250 mol/m^3. (The peak concentration was assumed to reach a plateau, because two successive concentrations, measured when 1.03 and 1.54 bed volumes had eluted, were nearly equal.) The most likely explanation for the asymptotic miminum and constant peak concentrations is that the plutonium in the bulk solution was in equilibrium with the adsorbed plutonium on the resin surfaces. Discharge plutonium concentrations should have approached equilibrium levels during this elution because the bulk solution was in contact with partially loaded resin over the entire length of the column.

The Ionac 641™ resin equilibrium saturation was estimated by assuming that equilibrium conditions existed for the first test elution. The ratio of the measured equilibrium solubilities for the wash and elution acids was equated to calculated equilibrium values at the measured acid molarities. This calculation yields a solubility constant, K_1, of 0.63 and a saturated equilibrium concentration of 80.6 kg/m^3 at the elution acid concentration of 250 mol/m^3.

It may be noted that the calculated equilibrium concentration is not the same as the maximum measured discharge concentration. One explanation for this apparent contradiction is that the presence of the resin beads does not change the effective cross- sectional flow area. If the cross-sectional area had changed to the fraction occupied by the bulk solution, then. the plutonium would have begun to discharge from the column well before one bed volume had eluted. The plutonium concentration spike does not appear until about one bed volume has eluted, however. This implies that the plutonium travels down the column at the superficial velocity. From a mass balance, it follows that the plutonium discharge concentration is diluted from the actual bulk solution concentration by the ratio of the cross-sectional area occupied by the bulk solution to the total cross-sectional area, i. e., the volume fraction occupied by the bulk solution. If the plutonium concentration in the bulk solution is in saturated equilibrium with the adsorbed resin, then

$$x = \varepsilon_0 x_{sat} \tag{46}$$

For $x = 26.3$ kg/m^3 and $x_{sat} = 80.6$ kg/m^3,

$$\varepsilon_0 = 0.326 \tag{47}$$

This value for the bulk solution volume fraction is slightly less than the value of 0.36 given for random close packing of uniformly sized spheres.[8]

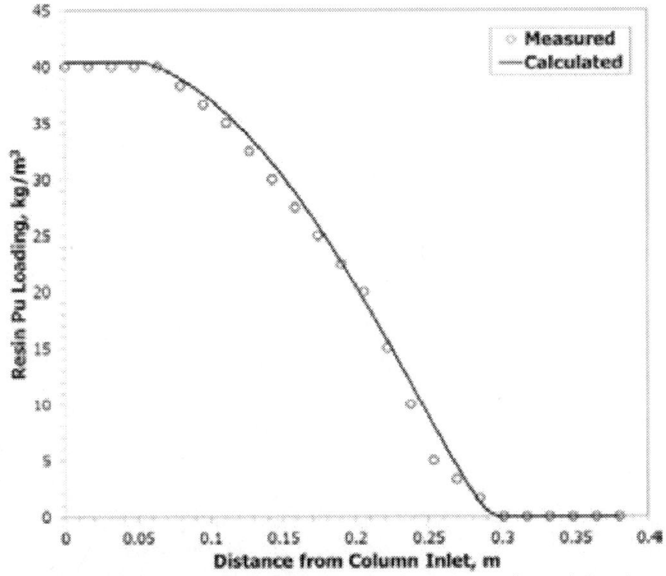

Figure 2. Comparison of Measured and Predicted Resin Loadings after 11 h Wash at 0.0005 m/s with 0.4 kg/m^3 Pu Solution

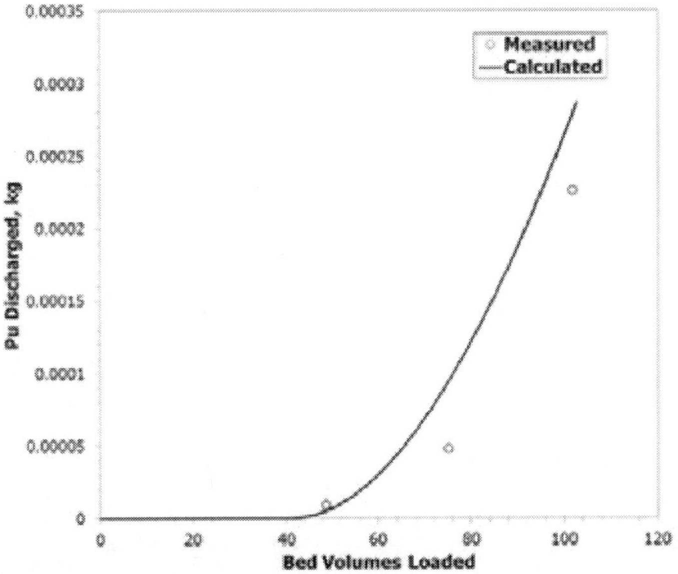

Figure 3. Comparison of Measured and Calculated Cumulative Pu Discharge for First Test Elution (Assumes 45 kg/m^3 Maximum Loading)

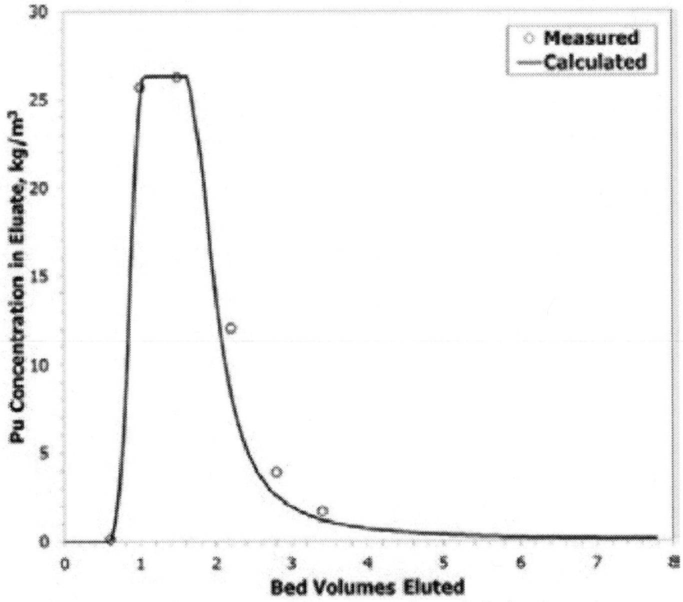

Figure 4. Comparison of Measured and Calculated Elution Profiles for First Test Elution

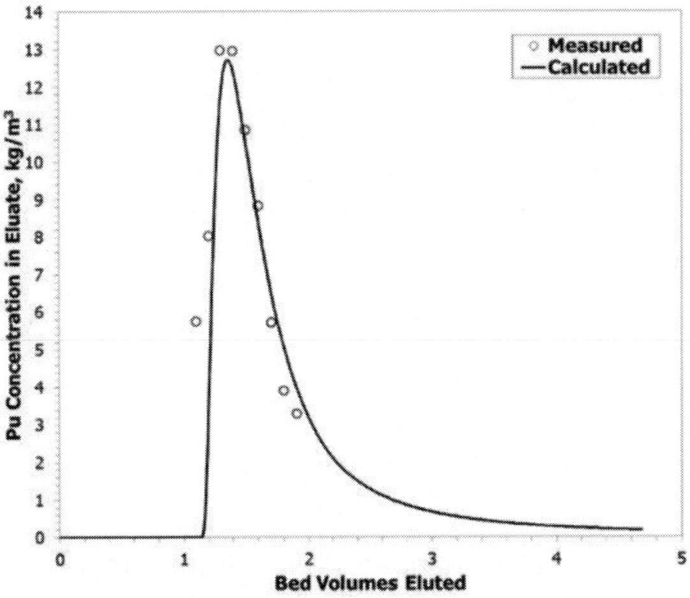

Figure 5. Comparison of Measured and Calculated Elution Profiles for Second Test Elution

Resin Bed Heat Transfer Model

Heat Transfer Model Development

A description of the energy balance used to model heat transfer in the resin bed during flow interruptions follows. The natural convection model includes two-dimensional mass transfer, but that model is not presented. During flow stoppages, if there is a significant plutonium load in the column, most of that plutonium will be adsorbed onto the resin beads, so to a good approximation, the plutonium concentration can be viewed as fixed, and mass transfer by natural convection is of only minor consequence.

The convective heat transfer equation includes a term for the decay heat, Q_{decay}. This equation is

$$\rho_{bed} c_{p,bed} \frac{\partial T}{\partial t} + \rho_f c_{p,f} v_z \frac{\partial T}{\partial z} + \rho_f c_{p,f} v_r \frac{\partial T}{\partial r}$$
$$= \left(k_{bed} + \rho_f c_{p,f} \alpha_z + \rho_f c_{p,f} D_{mix}\right) \frac{\partial^2 T}{\partial z^2} + \left(k_{bed} + \rho_f c_{p,f} \alpha_r\right) \frac{1}{r} \frac{\partial}{\partial r}\left(r \frac{\partial T}{\partial r}\right) + Q_{decay} C_{Pu} \tag{48}$$

The axial dispersion coefficients in the mass and heat transfer equations is given by Equation 4. The radial dispersion coefficient is about one-fifth as large.[9] Thus,

$$\alpha_r = 0.4 d_p v_z \tag{49}$$

For normal operating conditions (loading, washing, and elution), the superficial axial velocity is specified, and the superficial radial velocity is assumed to be zero. For flow stoppages, these velocities result from natural convection. Natural convection velocities are ignored during loading washing, and elution. At high feed flow rates, the omission of natural convection does not have a significant effect. At lower feed flow rates, this omission is conservative in that it results in higher resin bed temperatures than would be predicted if natural convection were included.

Velocity components for natural convection flow are calculated by combining the Ergun equation with axial and radial component momentum equations, as suggested by Stewart and Dona.[10] The axial velocity for natural convection is modeled using a form of the Ergun equation for laminar flow, which is[11,12]

$$v_z = \frac{d_p^2 \varepsilon_o^4 (\rho - \bar{\rho}) g}{200 (1 - \varepsilon_o)^2 \mu} \tag{50}$$

It may be noted that in Equation 54, $v_z \propto \frac{\varepsilon_o^4}{1 - \varepsilon_o^2}$, while in the Ergun equation, $v_z \propto \frac{\varepsilon_o^3}{1 - \varepsilon_o^2}$. The extra void fraction multiplier is added because the buoyant force applies only to the fraction of the total volume occupied by the bulk solution and not to the resin beads.

The Ergun equation was originally derived for unidirectional flow. To model the two-dimensional flow in the resin bed, separate equations are used for the axial and radial velocity components. These equations take the form

$$v_z = \frac{d_p^2 \varepsilon_o^4}{200 (1 - \varepsilon_o)^2 \mu} \left(-\frac{\partial P}{\partial z} + (\rho_f - \bar{\rho}_f) g \right) \tag{51}$$

and

$$v_r = \frac{d_p^{\,2} \varepsilon_o^{\,4}}{200\left(1-\varepsilon_o\right)^2 \mu}\left(-\frac{\partial P}{\partial r}\right) \tag{52}$$

The buoyancy term in the axial velocity equation contains a reference density, $\bar{\rho}$. At equilibrium, the reference density must be set so that there is no net flow upward or downward at any bed level. This implies that the reference density must equal the average fluid density at a given bed level. In other words,

$$\bar{\rho}_f = \frac{2}{R^2}\int_{r=0}^{R} r \rho_f dr \tag{53}$$

The pressure gradient terms must be eliminated from Equations 17 and 18 to solve for the velocity components. This is accomplished by taking cross derivatives. The resulting equation is

$$\frac{\partial v_r}{\partial z} - \frac{\partial v_z}{\partial r} + \frac{d_p^{\,2} \varepsilon_o^{\,4} g}{200\left(1-\varepsilon_o\right)^2 \mu}\frac{\partial \rho_f}{\partial r} = 0 \tag{54}$$

The radial velocity is calculated by combining the continuity relation,

$$\frac{1}{r}\frac{\partial r v_r}{\partial r} + \frac{\partial v_z}{\partial z} = 0 \tag{55}$$

with Equation 54.

The natural convection velocities defined by Equations 55 and 56 do not account for convection due to cross-sectional average axial density gradients. At the start of a flow stoppage, the solution in the resin bed is unstably stratified, with the higher plutonium concentrations and therefore the higher densities at the top of the resin bed. Preliminary calculations with these velocities showed that the solution remained unstably stratified even as equilibrium approached. Actually, significant unstable axial density variations would cause heavier fluid from above to mix with lighter fluid from below in regularly spaced regions called Bénard cells. A detailed model of Bénard cell convection was not attempted. Instead, mixing due to axial density gradients is calculated using an analogy to bubble columns. A dispersion model is used to describe mixing in bubble columns. The dispersion coefficient for bubble columns is defined in terms of the circulation velocity induced by the bubble motion, v_c, and the column diameter:[13]

$$D_{mix} = 0.24 d v_c \tag{56}$$

The same correlation is used to calculate the dispersion due to buoyancy-induced mixing in the resin bed. The circulation velocity is set equal to half that given by the Ergun equation and the pressure gradient based on the difference between the average densities for different levels:

$$v_c = \frac{d_p^{\,2} \varepsilon_o^{\,4}\left(\bar{\rho}_f\big|_z - \bar{\rho}_f\big|_{z+\Delta z}\right) g}{400\left(1-\varepsilon_o\right)^2 \mu} \tag{57}$$

The one-half factor appears because mixing occurs in both directions and can therefore occupy only half of the cross-sectional flow area in either direction. The use of cross-sectional average densities accounts for the axial pressure gradients neglected by Equation 51. That equation already incorporates local density variations in the radial direction, so local density gradients are not included.

In the model, the dispersion coefficient is calculated using a weighted distribution of circulation velocities for density differences at different levels. The distribution is based on the mixing height for the dispersion coefficient, which is 0.8 times the column diameter.[13] If it is assumed that the mixing intensity decreases exponentially with axial distance, these values for the mixing height and velocity give the following expression for the local mixing velocity:

$$v_{c,eff} = v_c \exp\left(-2\frac{|z - z_0|}{0.8d}\right) \tag{58}$$

where z_0 is the axial location at which the dispersion coefficient is evaluated. It can be confirmed that an integration of this equation with respect to axial displacement gives the mixing height.

<u>Boundary Conditions for Heat Transfer</u>

The solution of the energy equation required boundary conditions for thermal conduction and inlet conditions for natural convection. Mass balances for loading, washing, and elution provided the initial conditions when flow is interrupted.

Two conditions were specified for the LANL column, exposure to ambient air and cooling by a water jacket. For air exposure, the conduction of heat from the resin bed through the circumferential wall equals the combined rate of natural and forced convection and thermal radiation to the ambient air. For thermal radiation, a value of 0.6 was used for the surface emissivity ε. This emissivity is characteristic for steel surfaces and is a conservatively low estimate for glass surfaces.[14] Standard Nusselt number correlations were used to model natural convection[15] and forced convection.[16] For cooling by a water jacket, it was assumed that the wall temperature equals the water temperature. It was assumed that there is no heat transfer at the top and bottom of the column.

Inlet and outlet flow conditions were specified by assuming that the fluid exiting the resin bed due to natural convection becomes well-mixed in small volumes of free-standing liquid above and below the bed.

<u>Calculation Method for Natural Convection Calculations</u>

A forward time step, donor cell finite difference method was used to model the natural convection heat transfer. The flow terms in the heat transfer equations were expressed in terms of a stream function. Time step sizes of 0.02 s for operation at the specified feed flow rate and 1.0 s for natural convection during flow stoppages were used. Cell sizes were decreased until approximate convergence of temperature profiles and flow streamlines was obtained. It was determined that a grid of 48 axial cells by 24 radial cells gave an adequate approximation.

<u>Results of Natural Convection Heat Transfer Calculations</u>

Figure 6 depicts the effect of a flow stoppage on the maximum resin bed temperature for the 0.0746-m diameter by 0.341-m tall LANL column. This figure shows results for a case where the column would be inadvertently fed twice its capacity of 0.05 kg Pu-238. During a flow stoppage, the maximum column temperature would increase from 331 K to 358 K if the column were air cooled and from 318 K to 335 K if the column were cooled by a water jacket. Even for air cooling, the temperature remains well below the 388 K minimum temperature for initiation of a reaction excursion, so no water jacket is needed.

Figures 7 and 8 illustrate the temperature profile and the recirculation streamlines for natural convection for the air cooling case from Figure 6. The streamlines in Figure 8 demonstrate that there are two circulation cells, one in the top portion of the column and one in the bottom portion. Within both cells, the flow is upward near the column centerline and downward along the column wall. The top cell is slightly stronger than the bottom cell, so temperatures in the bottom of the column slightly exceed the temperatures in the top of the column.

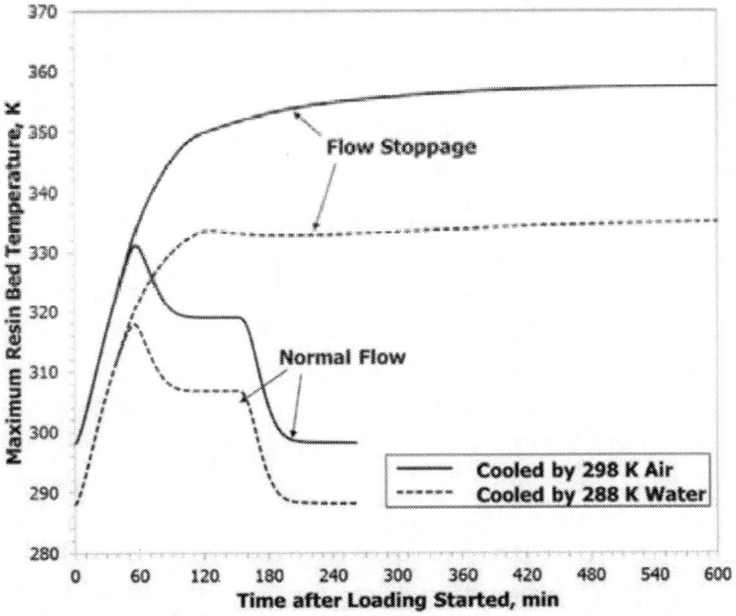

Figure 6. Maximum Resin Bed Temperatures for a Low Feed Flow Rate, High Feed Concentration, with Twice the Resin Capacity Fed

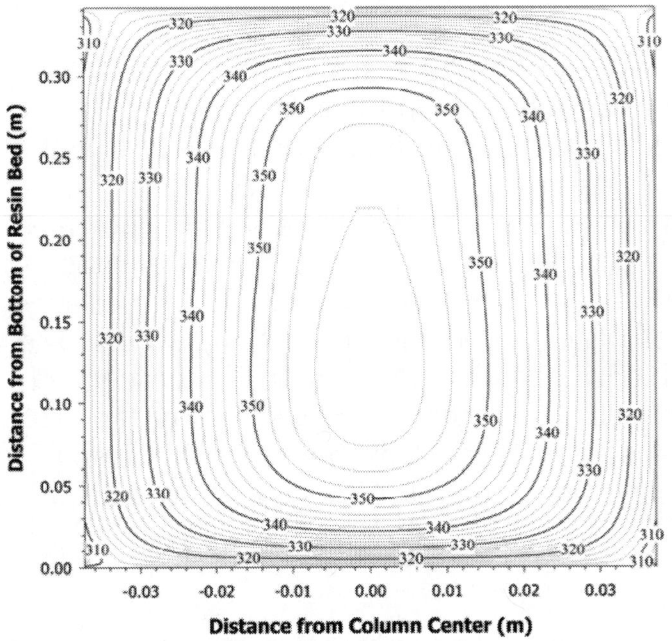

Figure 7. Temperature Distribution at Thermal Equilibrium after a Flow Stoppage for Low Feed Flow Rate, High Feed Concentration, Exposure to 298 K Air, and Twice the Resin Capacity Fed

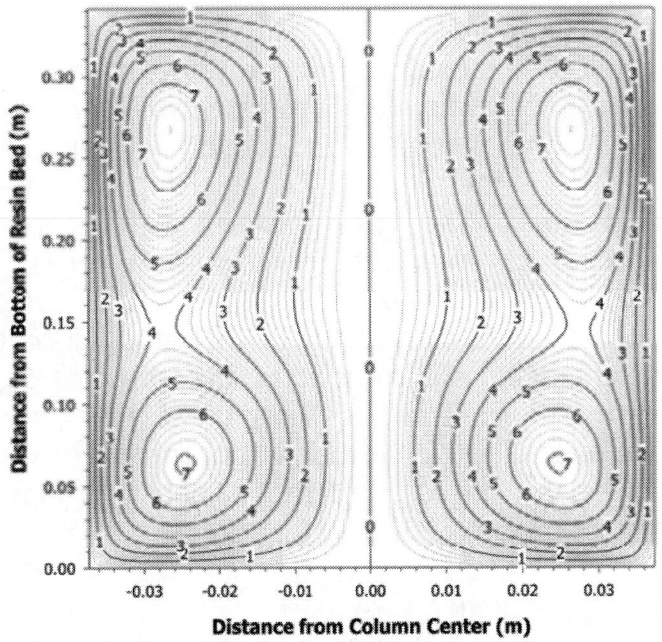

Figure 8. Flow Streamlines at Thermal Equilibrium after a Flow Stoppage for Low Feed Flow Rate, High Feed Concentration, Exposure to 298 K Air, and Twice the Resin Capacity Fed (Numbers on graph are values of the stream function in 1E-7 m^2/s.)

Resin Dose Calculations

Impact of Resin Dose on Ion Exchange Column Safety and Performance

Recently, the resin bed model was used to estimated doses to the resin for purification of plutonium in a 0.015-m diameter, 0.5-m tall column containing Reillex HPQ™ resin. The feed stream to the ion exchange column was to contain plutonium with a mixed isotope distribution, contaminated with the principal radioactive isotopes Am-243, Cm-244, Cs-137, and Sr-90. (The cesium and strontium are present as fission products.) The ion exchange column purifies the feed stream by preferentially adsorbing the plutonium. The distribution coefficient for the preferential adsorption of plutonium relative to americium and curium was estimated to be 165, based on laboratory tests.[17] The resin does not adsorb cesium or strontium to any significant extent.

During the ion exchange column operation, irradiation from adsorbed and non-adsorbed radionuclides decreases the exchange capacity of the resin and makes it more susceptible to attack by the nitric acid in the loading and washing solutions. At high temperatures, the nitric acid-resin decomposition reactions are exothermic and self-accelerating. These reactions pose a safety risk because they release product gases that can pressurize the ion exchange column.

In SRNL thermal degradation tests, rates of heating and pressurization did not change significantly for resin exposed to a 3E6 Gy dose but exhibited significant changes for a dose of 5E6 Gy. Previous tests conducted at Los Alamos National Laboratory (LANL) demonstrated that the plutonium adsorption capacity of the resin decreased by 10% or less for a dose of 3E6 Gy.[18] Based on these results, the recommended maximum radiation dose for the Reillex HPQ™ resin was set at 3E6 Gy.

Resin Dose Model

The total dose to the resin was calculated as the sum of the doses from individual radioactive isotopes. Doses are based on the residence time of each isotope in the column.

The dose analysis accounts for partial shielding of the resin from alpha radiation by the acid solution in the column. This shielding is evaluated in terms of the travel distance of an alpha particle in solution. The alpha particle travel distance is correlated in terms of the atomic number of the medium containing the alpha particle, the particle energy, and the absorber material density.[19] It has been determined that, if the alpha particles travel a uniform distance and lose energy at a constant rate along their travel paths, all particles generated within one travel path distance of a solid surface will transmit, on average, one-eighth of their energy to the solid.[20] Consequently, the resin beads are subject to alpha radiation only from solution within one travel path distance of their surfaces, and within that range, the beads only adsorb one-eighth of the total alpha energy. It is conservatively assumed that all beta and gamma radiation penetrates the beads and is deposited in the column.

Dose Calculation Results

Figure 9 depicts the dose profile at the end of the feeding and elution stage. At the end of elution, the total cumulative dose at the top of the column is 8170 Gy, well below the limiting dose of 3E6 Gy. Slightly less than half the maximum dose is from plutonium, and slightly more than half the maximum dose is from curium. The plutonium dose is primarily from the plutonium that adsorbs onto the resin, and the curium dose is mostly from curium that passed through the column during the feeding stage. Americium, cesium, and strontium contribute only slightly to the total dose.

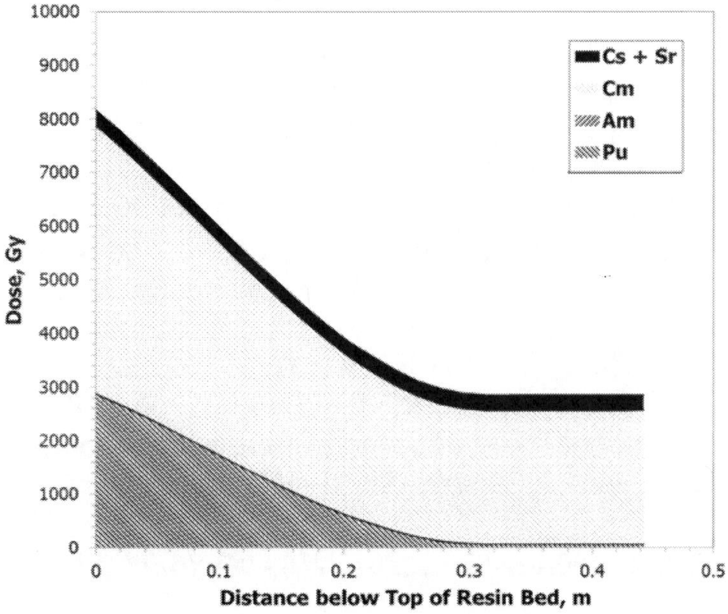

Figure 9. Cumulative Dose on Column after Elution

Conclusions

A finite element code has been developed to model mass and heat transfer in anion exchange columns used to extract plutonium. The model is applied to model unidirectional forced flow during normal column operations and natural convection during flow stoppages. Resin loading is modeled by combining a chemical equilibrium absorption relation with models for axial dispersion and solid phase diffusion. Elution is modeled by combining the axial dispersion model with a shrinking core diffusion model. Natural convection calculations are based on the Ergun relation for flow in a packed bed. The model treats the resin bed as a continuum so that a regular axisymmetric calculation grid can be used, without any fine discretization to accommodate boundary layers. As a consequence, the code is computationally efficient.

Calculations demonstrate that the code can accurately model loading profiles and elution transients for typical plutonium extraction operations. Examples are presented for the modeling of natural convection during a flow stoppage and for calculation of the radiolytic dose to the resin for an extraction cycle.

Acknowledgment

This manuscript has been authored by Savannah River Nuclear Solutions, LLC under Contract No. DE-AC09-08SR22470 with the U.S. Department of Energy. The United States Government retains and the publisher, by accepting the article for publication, acknowledges that the United States Government retains a non-exclusive, paid-up, irrevocable, world-wide license to publish or reproduce the published form of this manuscript, or allow others to do so, for United States Government purposes.

Partial funding for this work was provided by Los Alamos National Laboratory.

References

1. Laurinat, J.E., "Self-Heating of Pu-238 Anion Exchange Column (U)," Westinghouse Savannah River Company Report No. WSRC-TR-94-0166, March 1994.
2. Laurinat, J.E., "Thermal Analysis of LANL Ion Exchange Column," Westinghouse Savannah River Company Report No. WSRC-TR-99-00027, May 1999.
3. Pansoy-Hjelvik, M.E., J.Z. Nixon, J. Laurinat, J. Brock, G. Silver, M.A. Reimus, and K.B. Ramsey, "Process Parameters Optimization in Ion Exchange ^{238}Pu Aqueous Processing," Los Alamos National Laboratory Report No. LA-UR-00-1546, presented at Plutonium Futures – The Science Conference, Santa Fe, NM, July 10-13, 2000.
4. Laurinat, J.E., "Revised Thermal Analysis of LANL Ion Exchange Column," Westinghouse Savannah River Company Report No. WSRC-TR-2006-00123, April 2006.
5. Ryan, J.L. and E.J. Wheelwright, "The Recovery, Concentration, and Purification of Plutonium by Anion Exchange in Nitric Acid," General Electric-Hanford Atomic Products Operation Report HW-55893, January 1959.
6. Perry's Chemical Engineers' Handbook, 6th ed., D.W. Green, ed., McGraw-Hill, New York (1984), p. 16-27.
7. Vermeulen, T., "Theory for Irreversible and Constant-Pattern Solid Diffusion," Ind. Eng. Chem., 45(8), 1664-1670, 1953.
8. Dullen, F.A.L., Porous Media: Fluid Transport and Pore Structure, 2nd ed., Academic Press, San Diego, California (1992).
9. Ebach, E.A. and R.R. White, "Mixing of Fluids Flowing through Beds of Packed Solids," A.I.Ch.E. J., 4(2), 161-169.
10. Stewart, Jr., W.E. and C.L.G. Dona, "Free Convection in a Heat Generating Porous Medium in a Finite Vertical Cylinder," J. Heat Transfer, 110(2), 1988, 517-520.
11. Perry's Chemical Engineers' Handbook, 6th ed., D.W. Green, ed., McGraw-Hill, New York (1984), p. 5-54.
12. Leva, M., Fluidization, McGraw-Hill, New York (1959), p. 47.
13. Joshi, J.B., "Axial Mixing in Multiphase Contactors – A Unified Correlation," Trans. I. Chem. E., 58, 155-165, 1980.
14. Perry's Chemical Engineers' Handbook, 6th ed., D.W. Green, ed., McGraw Hill, New York (1984), p. 10-51.
15. Perry's Chemical Engineers' Handbook, 6th ed., D.W. Green, ed., McGraw Hill, New York (1984), p. 10-13.
16. Kreith, F. and W.Z. Black, Basic Heat Transfer, Harper and Row, New York (1980), p. 251.
17. Kyser III, E.A. and W.D. King, "HB-Line Anion Exchange Purification of AFS-2 Plutonium for MOX," SRNL Report SRNL-STI-2012-00233, Rev. 1, July 2012.
18. Marsh, S.F., "The Effects of In Situ Alpha-Particle Irradiations on Six Strong Base Anion Exchange Resins," LANL Report LA-12055, April 1991.
19. Friedlander, G., J.W. Kennedy, and J.M. Miller, Nuclear and Radiochemistry, 2nd ed., John Wiley & Sons, New York (1949), p. 95.
20. Laurinat, J.E., N.M. Askew, and S.J. Hensel, "Flammability Analysis for Actinide Oxides Packaged in 9975 Shipping Containers," Paper PVP2013-97478, Proceedings of the 2013 ASME Pressure Vessels & Piping Conference, Paris, France, July 14-18, 2013.

Effect of Cooling on Natural Circulation inside Heated Channel Representing Prismatic Modular Reactor Core

Authors:
Mahmoud M. Taha, Ibrahim, A. Said, Shoaib Usman, and Muthanna Al-Dahhan

Acknowledgement for the presenter: **Laith Sabri**

MISSOURI S&T

Missouri University of Science and Technology

AICHE17

Introduction

- Today's global energy market places many demands of power generation technologies. (Zohuri 2016)
- In 2000, the Generation IV International Forum (GIF) signed the GIF charter that categorized the nuclear power plants into four categories. (Chapin et al. 2004, Southworth et al. 2004)

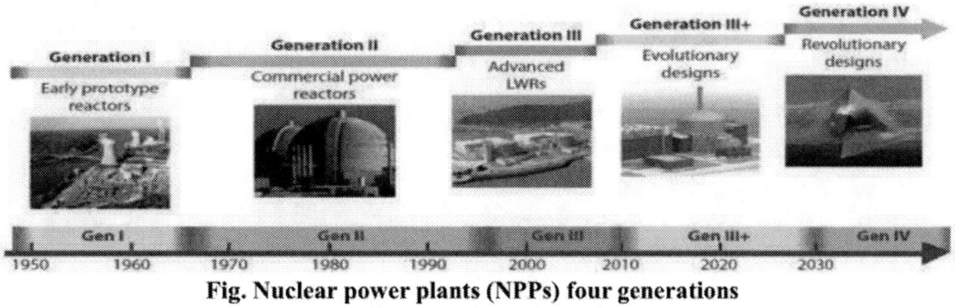

Fig. Nuclear power plants (NPPs) four generations

2

Prismatic Modular Reactor (PMRs)

Design concepts

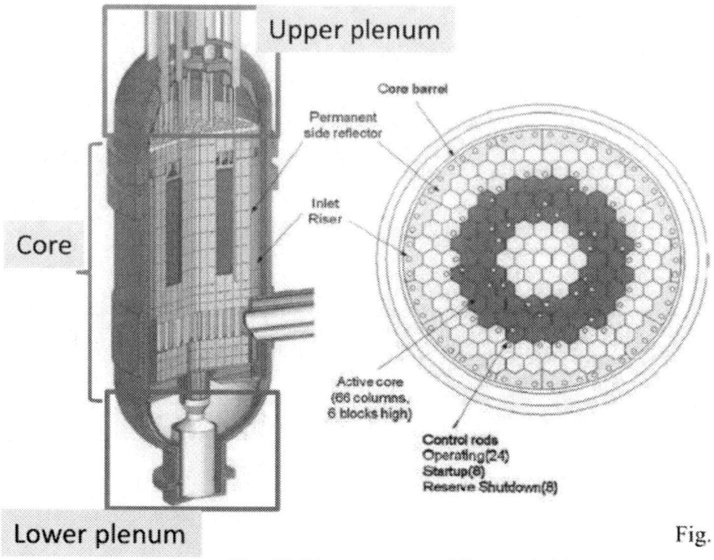

Fig. PMR components (Kim et al. 2011)

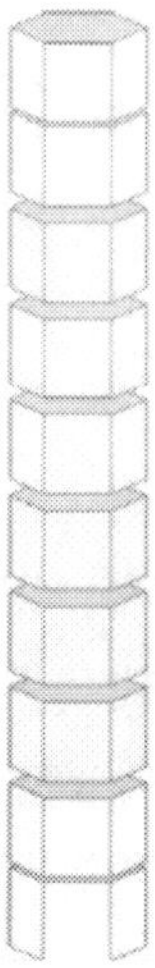

Fig. Hexagonal blocks

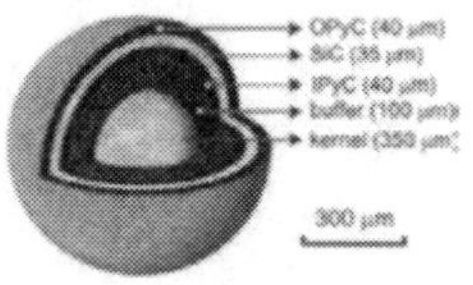

Fig. Fuel particles

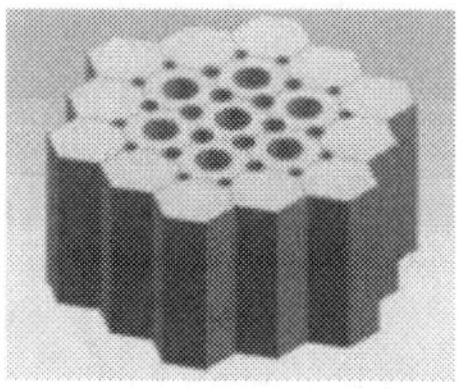

Fig. Fuel and flow channels

Prismatic Modular Reactor (PMRs)

Forced cooling - flow path

- Gas enters vessel through outer section of integrated duct.
- Flows downward through the core support structure.
- Gas flows upward between the outer and inner core barrels.
- Crosses upper plenum and flows downward through core coolant channels.
- Enters lower plenum and exits vessel through inner section of integrated duct.

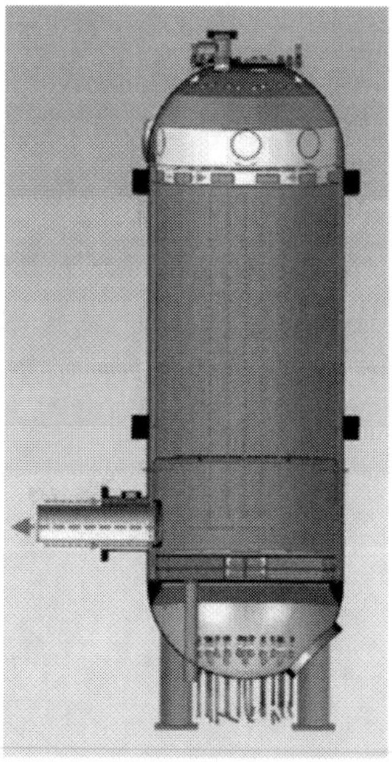

Prismatic Modular Reactor (PMRs)

Passive cooling - flow path

- In case of Loss of Flow Accidents (LOFA), forced cooling circulation system lost.
- Radial temperature gradients exist within the core as a result of decay heat.

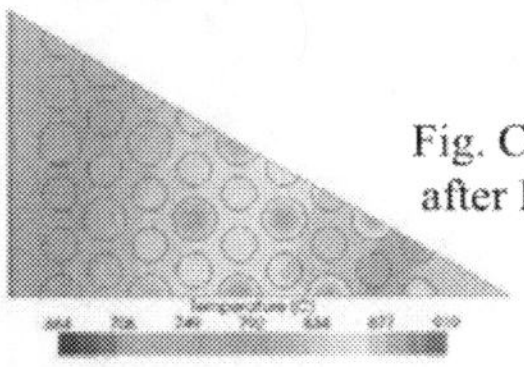

Fig. Core temperature gradient after LOFA (Tung et al. 2011)

- Buoyancy forces overcome the gravity and natural circulation initiated and flow direction is reversed.

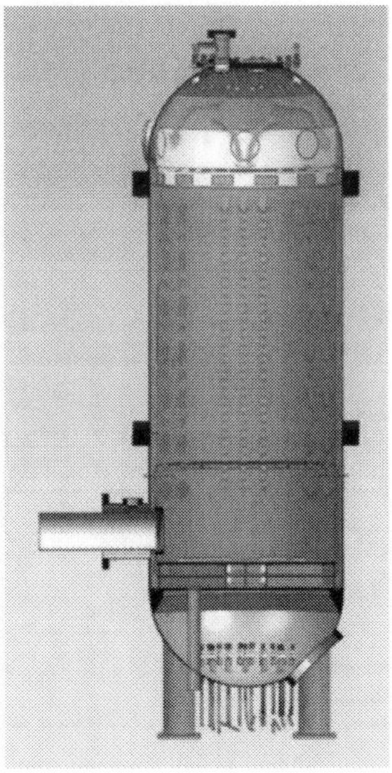

5

MISSOURI UNIVERSITY OF SCIENCE AND TECHNOLOGY

Prismatic Modular Reactor (PMRs)

- High concern directed to PMRs safety in general and to natural circulation passive safety system in particular after the Fukushima Daiichi accident power plant accident in 2011. (Zohuri 2016)
- In 2012, it was declared that any nuclear system will be licensed **ONLY** if it fulfills the requirements on Generation IV safety goals. (Gen-IV report, 2012)
- The U.S. Department of Energy (DOE) calling for proposals every year for experimental studies investigating intra-core natural circulation thermal hydraulic phenomena.

6

Design of Plenum-to-Plenum Facility (P2PF)

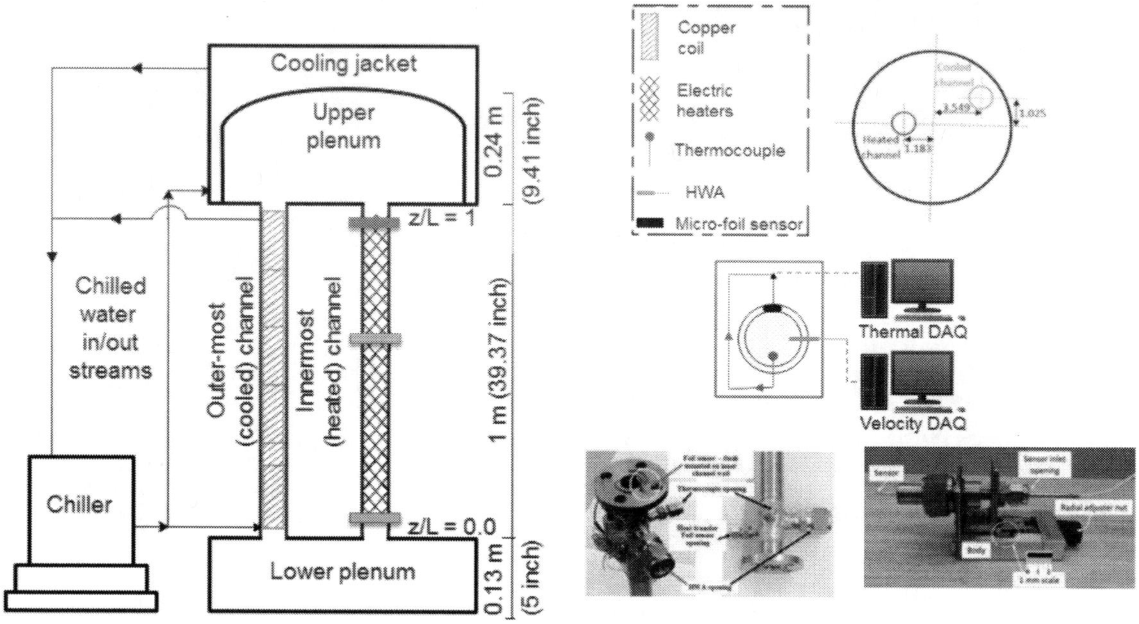

Findings and discussions

Temperature and velocity fields

- **Variables studied:**
  1. Chilled water temperature effect $(T = 5, 15, 25, and\ 35\ °C)$ on flow inside heated channel.

- **Other variables are kept constant:**
  1. Heating intensity at 700 W/m^2.
  2. Atmospheric pressure air used as the working fluid.
  3. All measurements are collected after reaching thermal steady state.

Findings and discussions

Temperature and velocity fields

- For better illustration of flow inside the channel, it is divided into two sections:
  1. Inlet section that expands from the leading edge (z/L = 0.0) to mid-section (z/L = 0.6) of the channel.
  2. Outlet section towards the channel exit (z/L = 1)

MISSOURI UNIVERSITY OF SCIENCE AND TECHNOLOGY

Findings and discussions

Temperature and velocity fields

- **Inlet section**

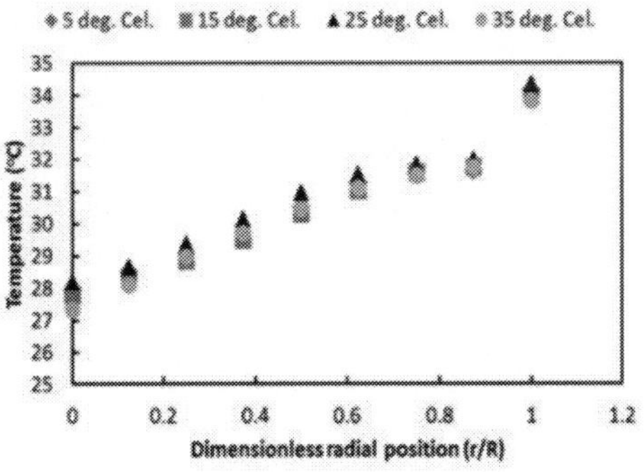

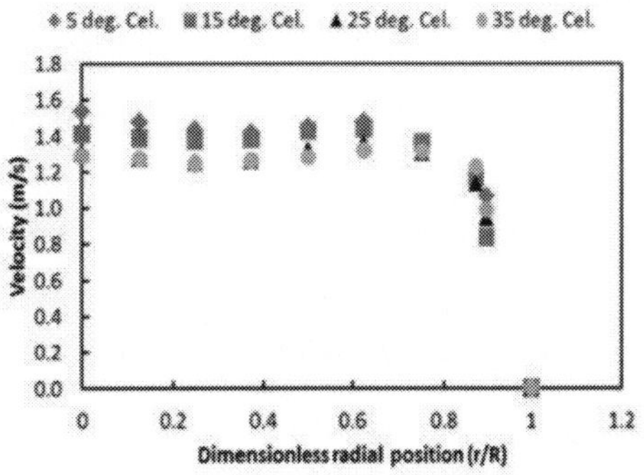

Findings and discussions

Temperature and velocity fields

- **Outlet section**

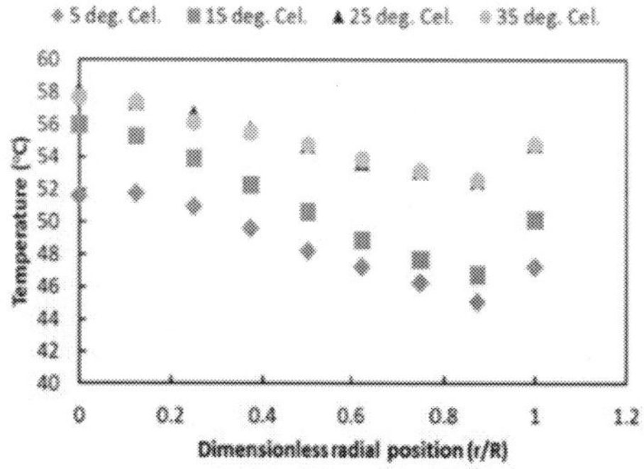

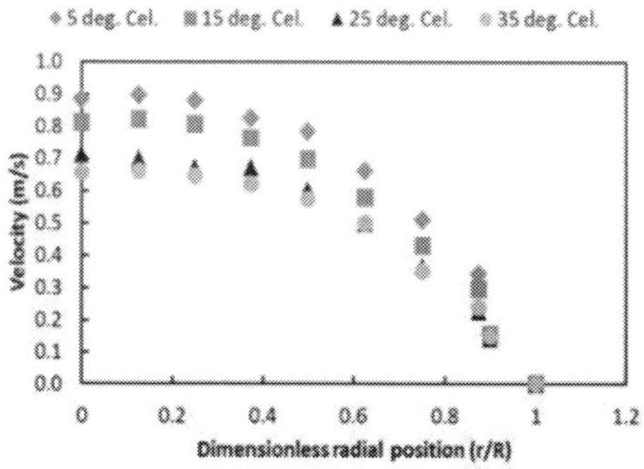

Findings and discussions

Flow field inside the heated channel

$$U_{rms}=\left(\frac{1}{N-1}\sum_{i=0.0}^{i=N}(U_i-U)^2\right)^{0.5}$$

- Turbulent intensities

$$I=\frac{U_{rms}}{U}$$

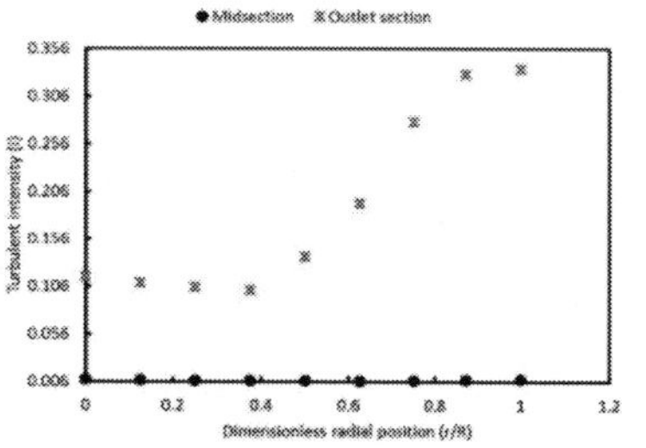

Fig. Radial distribution of turbulent intensity at different axial positions

Key observations:

- Turbulent intensity increase near channel wall at channel exit implies flow destabilization exists.

Findings and discussions

Flow field inside the heated channel

- At channel outlet
  - Amplification of turbulent intensity
  - Velocity atypical distribution

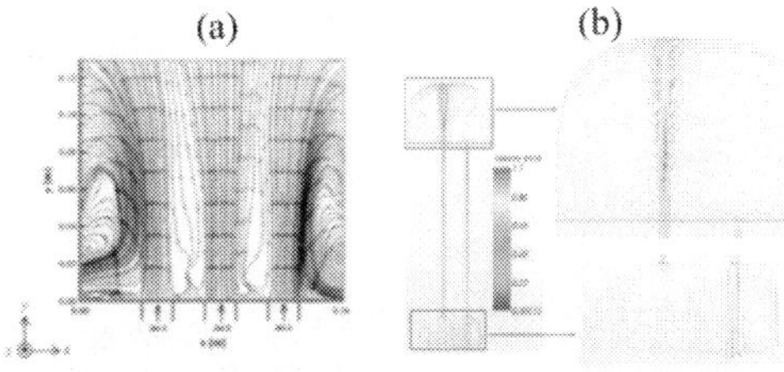

Fig. CFD results of upper plenum mixing (a) McVay et al. 2015 and (b) Kao et al. 2015

  - Upper plenum mixing. (Kao et al. 2015, McVay et al. 2015)
  - Fluid-to-wall heat transfer. (Said et al. 2017)
  - Heat losses by conduction through stainless steel flanges. (Kao et al. 2015)

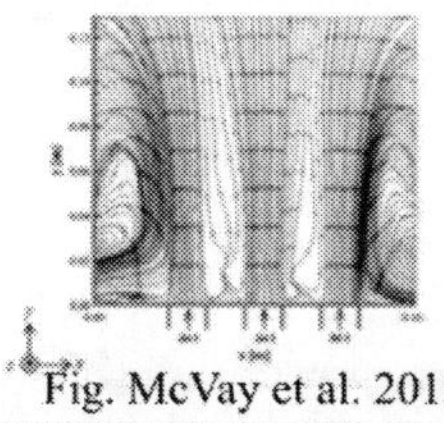

Fig. McVay et al. 2015

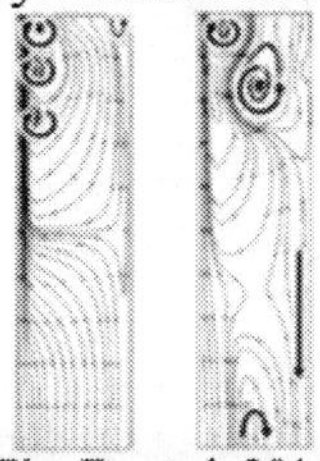

Fig. Fu et al. 2016

- Heated air exits the channel and cooled in the upper plenum.
- Cooled air is recirculated inside the heated channel and interact with the heated air near the channel wall.
- Flow interaction and mixing enhances heat transfer and hence causes a reduction in surface temperature and consequently velocity near the wall of the channel.
- Therefore the velocity peak is shifted towards the centerline of the channel.

13

Findings and discussions

Flow interaction at heated channel outlet

- Chilled water temperature effect ($T = 5, 15, 25, and\ 35$ °C) on flow inside heated channel

Table. Air temp. change with chilled water temp. change

Axial position	Low heating	High heating
Outlet (z/L = 0.96)	12.4%	6.4%
Mid-channel (z/L = 0.59)	1.4%	1.4%
Inlet (z/L = 0.04)	0.5%	0.4%

Table. Air velocity change with chilled water temp. change

Axial position	Low heating	High heating
Outlet (z/L = 0.96)	30%	45%
Mid-channel (z/L = 0.59)	25%	21%
Inlet (z/L = 0.04)	10%	7%

Key observations:

- Heated air exiting the channel is influenced by the degree of cooling.
- Interaction exits between cooled air recirculated from the upper plenum and heated air upcoming inside the channel.

14

Summary and remarks

- Separate effects test facility (P2PF) designed and developed with a representative geometry of PMRs core for investigation plenum-to-plenum thermal hydraulics phenomena under different natural circulation intensities.
- Atypical temperature and velocity distribution observed at channel outlet as a result of flow interaction and reversal at this section.
- Heated air exiting the channel is influenced by the degree of cooling.
- Interaction exits between cooled air recirculated from the upper plenum and heated air upcoming inside the channel.

MISSOURI UNIVERSITY OF SCIENCE AND TECHNOLOGY

THANKS ...

Questions ?!!

m.moharam@mst.edu
aldahhanm@mst.edu

Savannah River National Laboratory™

OPERATED BY SAVANNAH RIVER NUCLEAR SOLUTIONS

We put science to work.™

Impact of Impurities and Degradation Products During Chemical Processing of High Level Waste in the Defense Waste Processing Facility

Dan Lambert, Wesley H. Woodham, Matthew S. Williams, J. David Newell, Jack Zamecnik

Session 477a, Advances in Nuclear Waste Processing and Management

November 1, 2017

SRNL-STI-2017-00221

Savannah River NUCLEAR SOLUTIONS

Outline

- **Background**
- **Flammability Issues**
  - Hydrogen Generation
  - Ammonia Generation
  - Antifoam Degradation
  - Thermolysis
- **Conclusions/Recommendations**

Savannah River National Laboratory™ We put science to work.™

Savannah River Site

- The Savannah River National Lab is located at the Savannah River Site
- Located at the site are forty-three ~1 million gallon tanks filled with high-level radioactive waste (8 have been closed/grouted)
- This is considered the worst environmental hazard in South Carolina

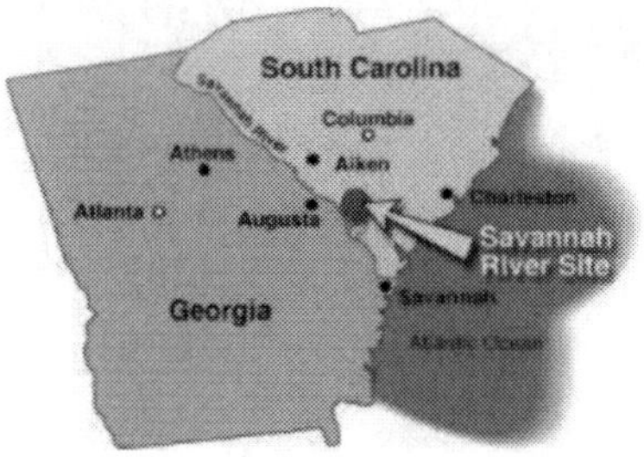

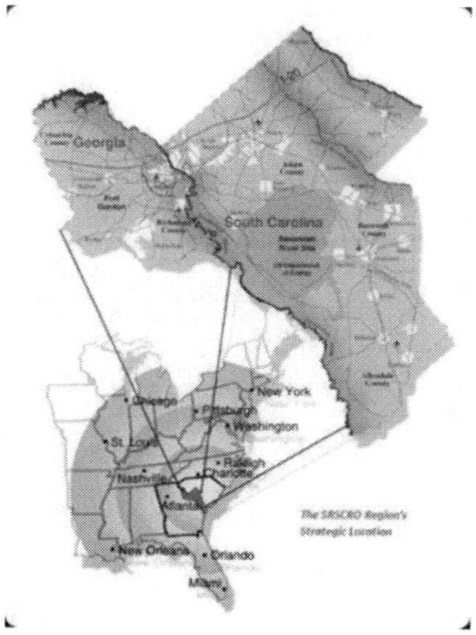

Savannah River National Laboratory We put science to work. 3

DWPF Process Overview
Formic Acid Flowsheet

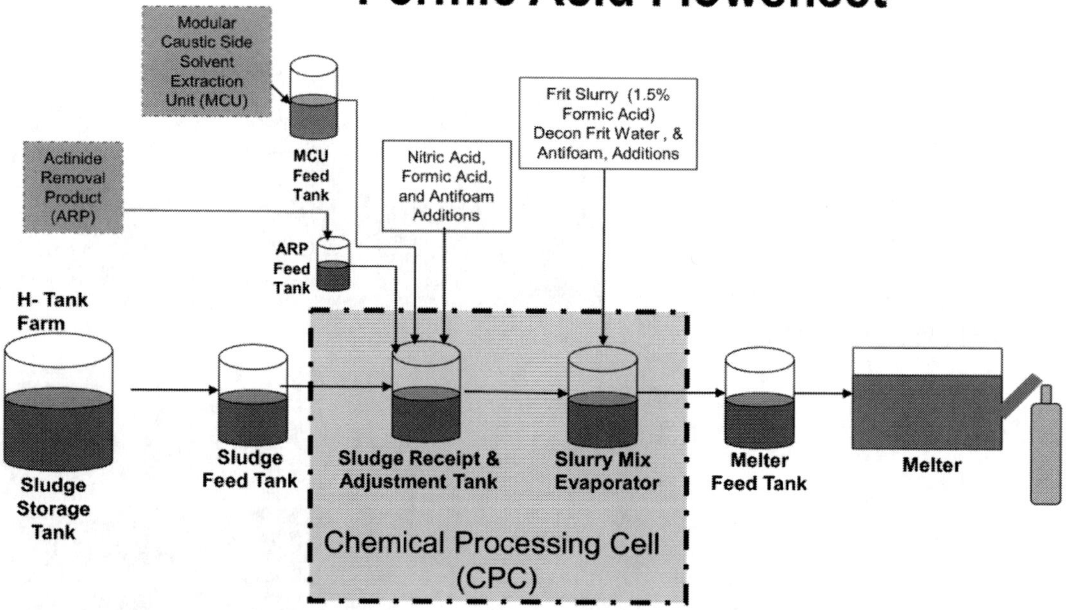

Savannah River National Laboratory We put science to work.

DWPF Process Overview
Glycolic Acid Flowsheet

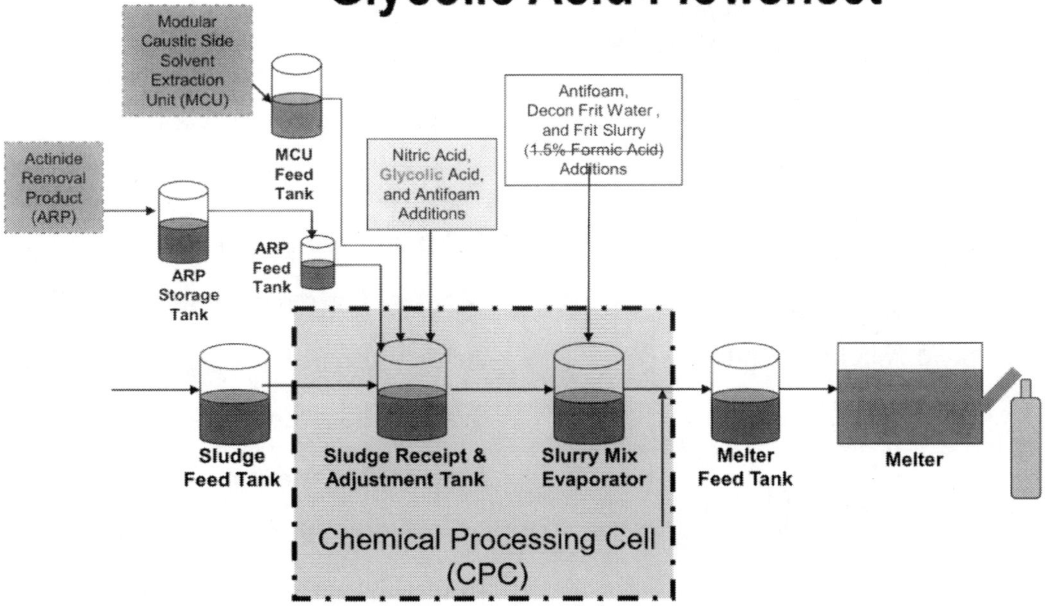

Why Change DWPF CPC Flowsheet?

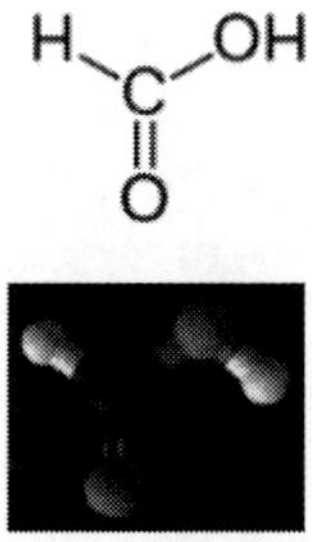

Formic Acid

- **Formic acid not only reduces mercury but also produces hydrogen**
- **If catalytic H_2 can be eliminated**
  - Air purge can be decreased
  - Gas chromatographs not safety significant
- **Without formic acid, rheology is thinner, pH more stable, complexes metals**
- **Glycolic acid does not produce hydrogen**

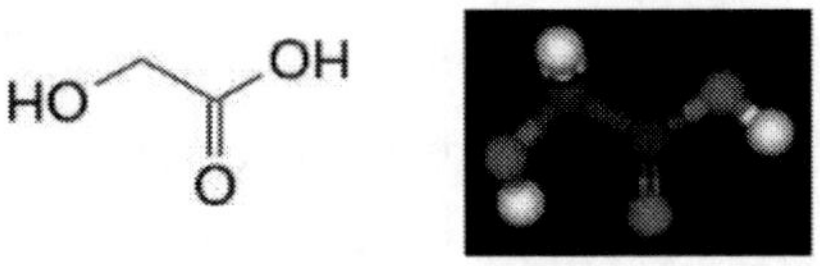

Glycolic Acid

What is Glycolic Acid (hydroxyacetic acid)?

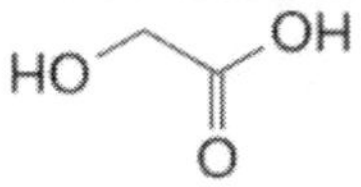

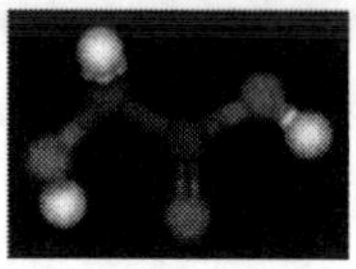

Glycolic Acid

- Smallest α-hydroxy acid
- Colorless, odorless, and hygroscopic crystalline solid
- Highly soluble in water
- A few million kilograms are produced annually

<u>**Commercial Uses**</u>

Facial Peel

- **Physicians 20-70 wt%**
- **Home use 10-20 wt%**

Soap and scum removal

Mineral scale removal

Acidity control

Water treatment

Boiler scale removal

- **2 moles Glycolic:mol Formic**

Bioabsorbed polymer sutures

DWPF Flammability Issues

- **Noble metals catalyze the decomposition of formic acid to hydrogen**

  $$HCOOH \rightarrow H_2 + CO_2$$

  - An air purge along with interlocks are used to ensure a flammable mixture of hydrogen and air is never formed
- **The strip effluent (contains ^{137}Cs) also contains about 87 ppm entrained solvent (primarily isopar L)**
  - The strip effluent is added slowly to minimize the impact
- **The antifoam degrades to three flammable degradation products**
  - The antifoam addition amount and timing between antifoam additions is controlled
- **Ammonia is generated during chemical processing**
  - Ammonia scrubbers are used to remove ammonia and prevent ammonium nitrate deposits

Flammable Issue: Hydrogen Generation

- **New flowsheet developed to reduce hydrogen generation**
- **Completed 100 experiments with glycolic acid**
- **Hydrogen generation is extremely low**
  - Very little formate is present (formic acid is impurity in glycolic acid) in runs with noble metals and mercury
  - SRAT purge has been reduced from 230 to 93.7 scfm
    - *Can't be lowered because of strip effluent isopar and antifoam degradation products*
  - SME cycle often has highest H_2 generation, purge remains at 74 scfm
  - In most runs hydrogen is detected only in SME cycle

Typical Hydrogen Profile

- **Very Low H2 generation in SRAT in Scaled Demo Runs GN70-79**

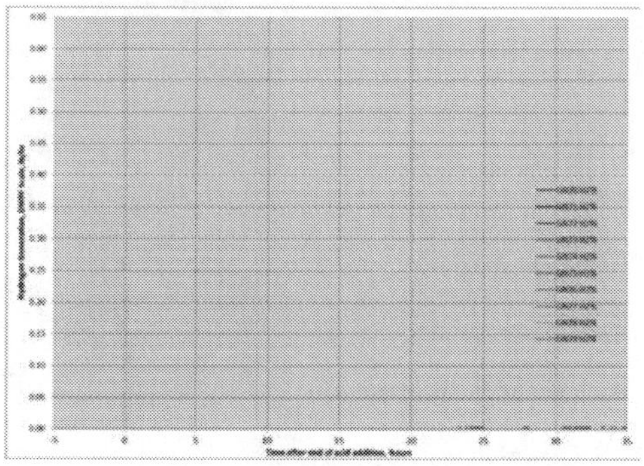

- **Zoom in 65x**

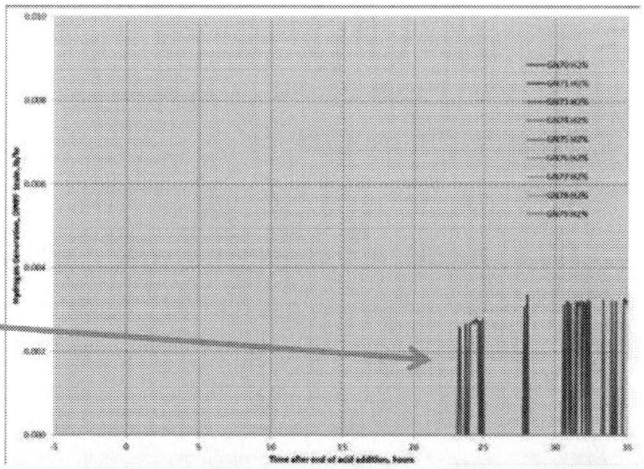

Flammable Issue: Ammonia Generation

- **Ammonia is generated from stepwise reduction of nitrate to ammonia**
  - During chemical processing, pH rises to ~10 in nitric-formic acid flowsheet
  - Equilibrium shifts from NH_4^+ to NH_3
  - Ammonia gas is released from slurry
  - Ammonia reacts with nitric acid to form ammonium nitrate

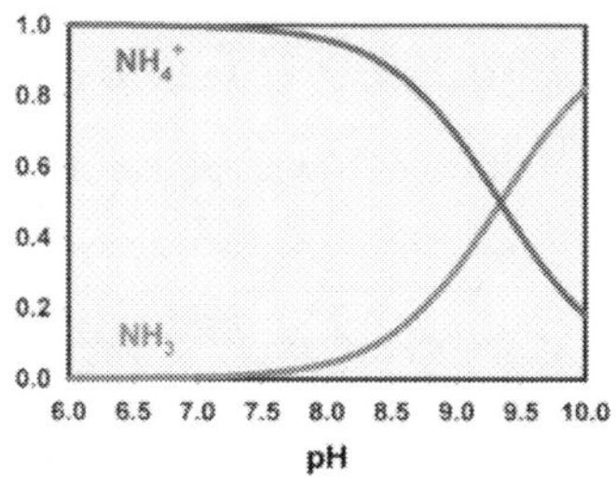

- **NO_2 results from decomposition of nitrite**
- **NO_2 is scrubber by water vapor and produces nitric acid**
  - Ammonia nitrate is an explosion hazard

Flammable Issue: Ammonia Generation

- **Very little ammonia is generated in GN Flowsheet runs**
  - No detected ammonia in offgas
  - Trace NH_4^+ detected in ammonia scrubber solution (35 mg/L vs 18,400 NO_3^-)
- **Low pH of SRAT product pushes equilibrium to ammonium**
  - $NH_3 + H^+ \leftrightarrows NH_4^+$
  - Ammonia would not be released even if produced
- **Glycolic Flowsheet**
  - NH_3 would stay in melter feed and be fed to melter
  - Ammonia scrubbers not needed for NH_3 removal
- **Formic Flowsheet:**
  - NH_3 would be produced in SRAT/SME, released to SME offgas
  - NH_3 would be removed by SRAT/SME scrubbers by SMECT liquid
  - NH_3 would be released in RCT with caustic addition or returned to tank far
  - NH_3 would be removed by RCT scrubber by SMECT liquid
  - In SE to SME runs, 2.5% of nitrate was converted to ammonia

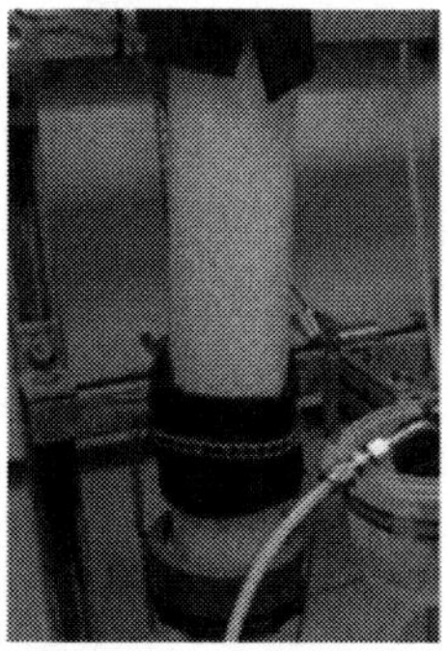

Flammable Issue: Antifoam 747 Degradation

- **Gas is generated during chemical processing**
  - N_xO_x generated due to nitrite destruction
  - CO_2 generate from carbonate destruction
  - Water vapor generated during boiling
- **Small insoluble solids stabilize foam and trap gases**
- **Foam generated during processing could cause a foam-over and lead to contamination, a complete facility shutdown, and the loss of millions of dollars**
- **Antifoam IIT747 implemented in DWPF in 2001 to reduce foaming**
- **Dozens of foam-overs have occurred, stopping processing in DWPF**

Flammable Issue: Antifoam 747 Degradation

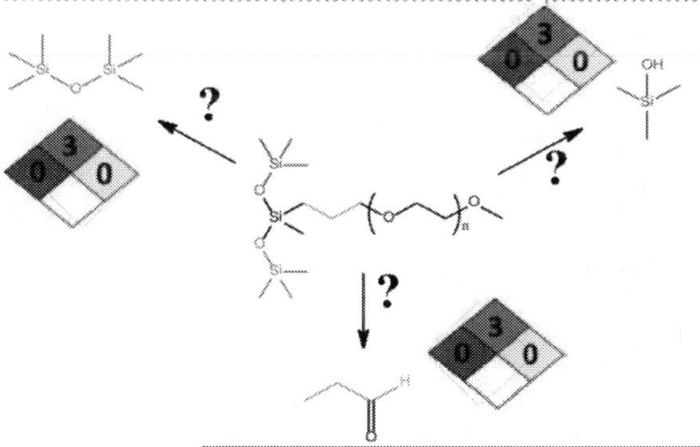

- **May 2015 – A Potential Inadequacy Safety Acquisition (PISA) was issued due to flammable antifoam degradation products which stopped all processing**

Compound	Formula	Structure	Molar Mass, g/mol	Solubility	Lower Flammability Limit, vol %	Boiling Point
Hexamethyl disiloxane (HMDSO)	$C_6H_{18}OSi_2$	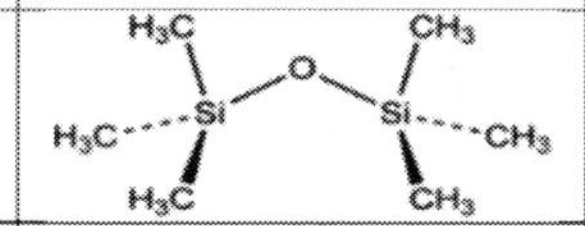	162.38	930.7 ppb	0.5-0.8	100 °C
Trimethyl silanol (TMS)	$C_3H_{10}OSi$	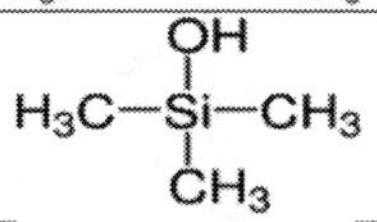	90.20	0.995 g/L	0.5-0.6	99 °C
Propanal	C_3H_6O	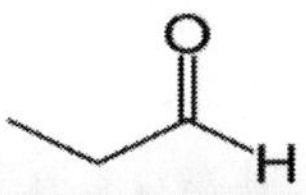	58.08	200 g/L	2.6	49 °C

Why does antifoam degrade so fast?

- **Antifoam 747 is most stable at pH 6.5-7.5**
- **Nitric-formic acid flowsheet**
  - Initial pH is ~12
  - pH drops to 4 by end of acid addition
  - Due to hydrogen generation and consumption of formic acid, pH rebounds to 10
- **Nitric-glycolic acid flowsheet**
  - Initial pH is ~11
  - pH drops to 4 by end of acid addition
  - pH rises slowly to about 5 by completion of processing
- **At initiation of boiling**
  - Gas generation is maximum (foaminess maximum)
  - Liquid boiling surface pH is <0 due to scrubbing of NO_2 as nitric acid

Savannah River National Laboratory We put science to work 15

HMDSO Flammability Concern

- **Hexamethyldisiloxane (HMDSO) insoluble antifoam degradation product**
- **Antifoam degrades at low and high pH**
- **Maximum HMDSO is at initiation of boiling**
  - Condensate very acidic
  - Peak temperature
- Short term fix
  - Add undiluted antifoam
  - Reduce antifoam addition to 1/6th
  - Add antifoam more frequently

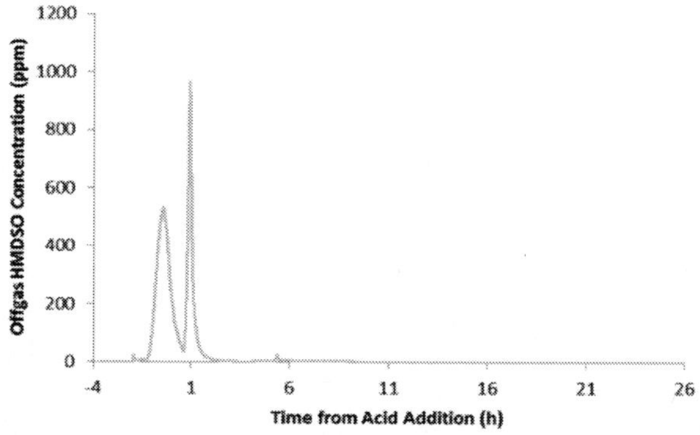

Savannah River National Laboratory We put science to work 16

Long Term Fix

- **New antifoam**
  - Hydrolysis resistant antifoam
  - Antifoam that doesn't produce flammable degradation products
- **Change processing to minimize antifoam decomposition**
  - Acidify slurry prior to boiling
  - Complete nitrite destruction prior to boiling
- **Mechanical fix**
  - Spray water to break foam
  - Install bladed in vapor space to break foam
- **Nonantifoam Chemical fix**
  - Flocculate solids
  - Large solids too big to stabilize foam

Flammability Issue: Thermolysis

- **Thermolysis or thermal decomposition of organic impurities leads to hydrogen generation**
- **This can happen in high level waste evaporators, in waste tanks, and in any downstream processing facility**
- **The 3 main organic impurities in the glycolic acid flowsheet are glycolate, formate and oxalate**
- **Formate and oxalate do not produce significant hydrogen via thermolysis, but glycolate does**
- **Although thermolysis is a slow reaction compared to catalyzed formic acid decomposition, it can be an issue in tanks and processing vessels with lower purges and higher temperatures**

Impact of Thermolysis

- **There is virtually no glycolate from chemical processing in DWPF due to glycolate's low vapor pressure**
  - Unless there is a foamover
- **In the melter, entrainment leads to glycolate in the condensate**
  - Higher during bubbled operations
- **Significant hydrogen peak at boiling**
- **How to minimize**
  - Improve antifoam
  - More gentle processing
  - Don't bubble melter unless needed
- **Testing ongoing to determine H_2 generation for glycolate**

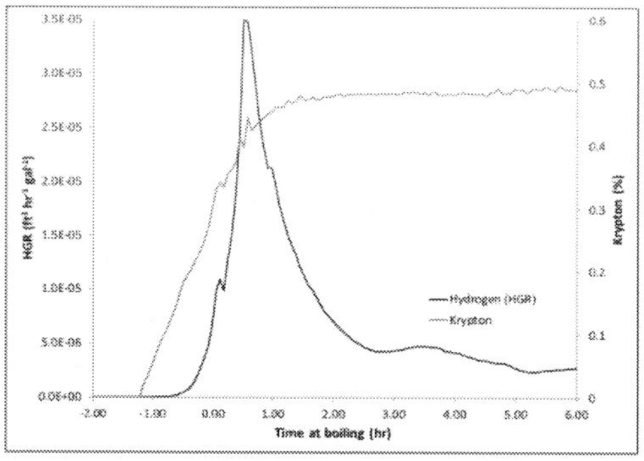

Conclusions

- **The nitric-glycolic acid flowsheet will**
  - Virtually eliminate hydrogen generation
  - Virtually eliminate ammonia generation
- **The current antifoam 747 should be replaced**
- **Determine H2 generation from glycolate thermolysis through testing**

Recommendations

- **Implement nitric-glycolic acid flowsheet in DWPF**
- **Develop hydrolysis resistant antifoam**
  - Processing changes may further reduce antifoam degradation
- **Complete testing to measure H2 generation from glycolate thermolysis**
  - Modify processing to minimize glycolate concentration in condensate

Acknowledgements

- **A large team has been involved in developing this flowsheet**
  - **SRNL researchers:** Mike Stone, Jack Zamecnik, Brad Pickenheim, Chris Martino, David Newell, Wes Woodham, Tom White, Whitney Riley, Frances Williams, Matt Williams, John Pareizs, John Mickalonis, Duane Adamson, et al
  - **SRNL Technicians:** Vickie Williams, Phyllis Workman, Jon Duvall, David Healy, Kim Wyszynski, Courtney Burkhalter, Madison Caldwell, Shielded Cells Technicians, et al
  - **SRR:** Terri Fellinger, Bill Holtzscheiter, Spencer Isom, Mason Clark, et al
  - **Interns:** Clayton Brandenburg, Michelle Luther, et al
- **Work was funded by the Department of Energy/Environmental Mgt**

Questions

Assessment of Dead End Filtration as a Screening Tool for LAWPS

John Geeting
Reid Peterson
Jarrod Allred
Renee Russell

TOPICS

- Comparison of CFF to DEF
- Apparatus
- AP-105 CUF Testing Results
- AP-105 DEF Testing Results
- Summary of CUF and DEF Simulant Results
- Conclusions

Problem Statement

Millions of gallons of radioactive waste to be treated at Hanford

Hanford tank waste treatment will be initiated by pretreating low activity waste in a facility known as the low activity waste pretreatment system (LAWPS)

The flowsheet integration team proposed that batch filtration tests using small quantities of feed be used to "qualify" each feed campaign prior to delivery to LAWPS.

To that end, testing was conducted to assess this proposal and involved:

- 1. Development of dead end filtration apparatus for use in the hot cell with radioactive feeds.
- 2. Evaluation of dead end filtration relative to cross flow filtration (CFF) with tank waste and simulants

Pacific Northwest NATIONAL LABORATORY

3

Proudly Operated by Battelle *Since 1965*

Advantages of DEF

- 1. Less feed
- 2. Simpler equipment
  - DEF is commercial off the shelf
  - CUF has many custom parts
  - DEF only requires a balance to calibrate inside the hotcell
  - CUF has Coriolis, magnetic flowmeter and RTDs in cell that have to be replaced on yearly cycle.
  - Other necessary services for CUF include compressed air and chilled water
- 3. Smaller footprint in hot cell
- 4. Lighter

Pacific Northwest NATIONAL LABORATORY

4

Proudly Operated by Battelle *Since 1965*

Basic Comparison of DEF to CFF

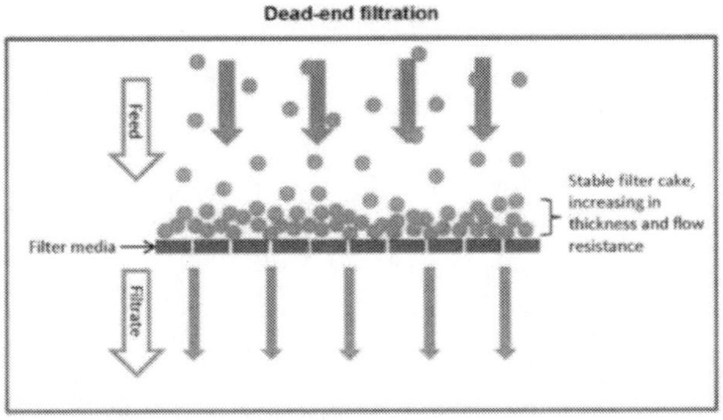

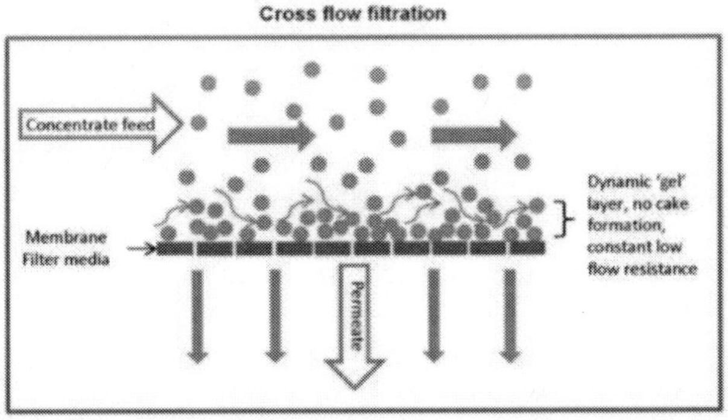

Reasoning as to why DEF might be used as a screening tool

- Assumption: LAWPS will only treat supernatant with very low insoluble solids content.
- $R_{total}=R_{media}+R_{fouling}+R_{cake}$
  - R_{media} – approximately the same between DEF and CFF
  - R_{cake} for DEF can grow very large. If small amount of solids then DEF and CFF may have similar resistance.
  - $R_{fouling}$-Are they similar?
- The hope: filtration using CFF and DEF may be sufficiently similar for screening purposes

CUF Piping Diagram

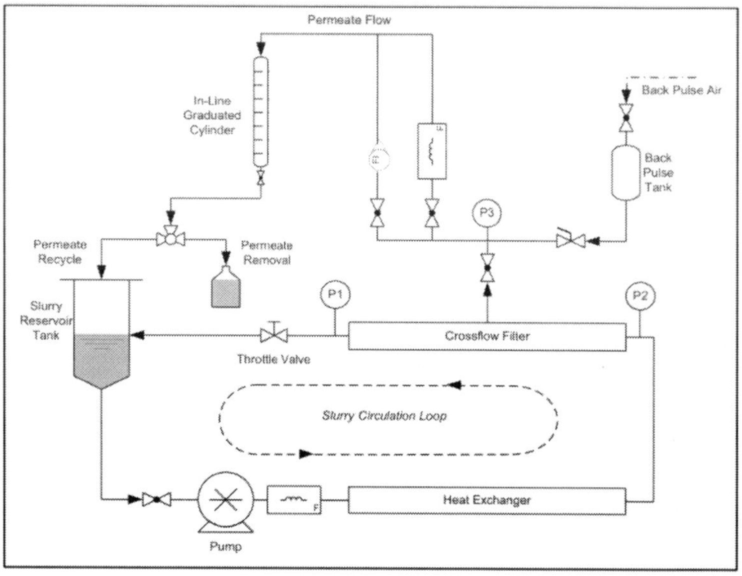

- Closed Loop Recirculation for Slurry
  - Reservoir Tank
  - Pump
  - Heat Exchanger
  - Filter
  - Throttle Valve
  - Magnetic Flow Meter
  - Pressure Gauges
- Permeate Manifold
  - Pressure Gauge
  - Mass Flow Meter
  - In-line Graduated Cylinder
  - Back-Pulse Chamber

The CUF

- Filter element
  - Variable: 0.5 to 2.0 ft
  - 0.1 micron rating Mott element
- System Volume
  - Maximum *4.1 Liters*
  - Operating Minimum *1.2 Liters*
  - Startup Minimum *1.5 Liters*
- Pump
  - Rotary Lobe *Up to 12 GPM*
- Instrumentation
  - Pressures *0 – 100 psig*
  - Temperature *0 – 100° C*
  - Slurry Flow Rate *Up to 15 GPM*
  - Permeate Flow Rate *Up to 30 lph*

8

Pacific Northwest NATIONAL LABORATORY

Proudly Operated by **Battelle** *Since 1965*

The DEF

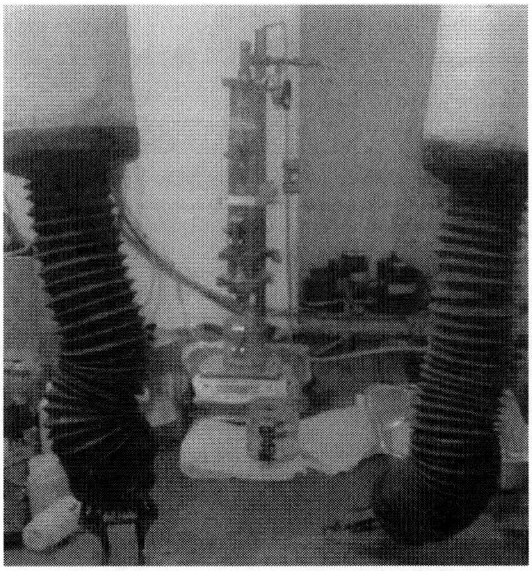

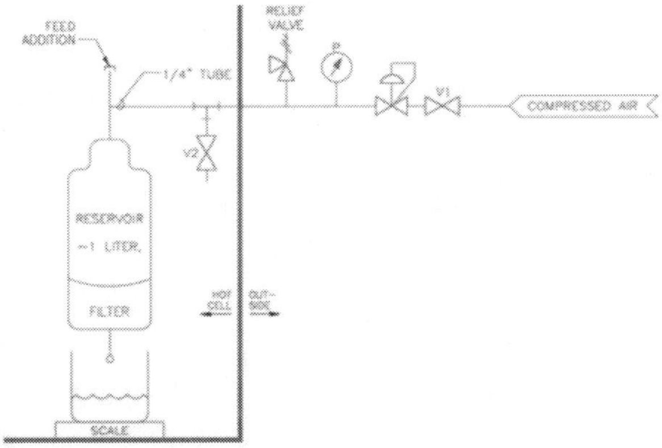

Comparison between DEF and CUF

Parameter	CUF	DEF
Filter	0.1 μm sintered SST	0.1 μm sintered SST
Manufacturer	Mott	Mott
Filter thickness (in)	1/16	1/16
Filter shape	Tubular (1/2-in ID)	Flat plate
Filter Area (ft^2)	0.065	0.031
Volume of feed (L) (nominal)	4	0.75
Volume/Area (cm)	66	26
Filtration Method	Cross-flow	Dead-end
Ease of construction	Difficult –many custom parts	Easy- generally COTS
M&TE Replacement	More difficult and expensive	Easy- outside hotcell
Footprint	Uses entire hotcell	Uses small portion of hotcell and can be easily moved

Filtration with AP-105 Tank Waste– completed twice

- Load feed into CUF
  - Diluted to 5.6 M Na
- Operate under recycle for ~ 28 hours
- Dewater to produce product for IX - total of 5 L produced
- Dead end filter balance – approximately 1.5 L (dewatered with varying pressure and constant pressure)
- CUF operated at 14.7 ft/s and various transmembrane pressure
- Temperature at 25C

Tank Waste Composition

	mg/L	Molarity
Na	196,000	8.53
Al	20,750	0.769
K	5525	0.141
S	1810	0.056
P	733.5	0.024
Cr	505	0.010
W	147.5	0.001
Mo	83.5	0.001
B	69.85	0.006
Ni	45.45	0.001
Viscosity	4.67	cP
Density	1.4	kg/liter
137Cs	208	µCi/mL
pH	13	
TOC	0.3	wt%

Pacific Northwest NATIONAL LABORATORY

11

Proudly Operated by Battelle *Since 1965*

CUF- Recycle Test

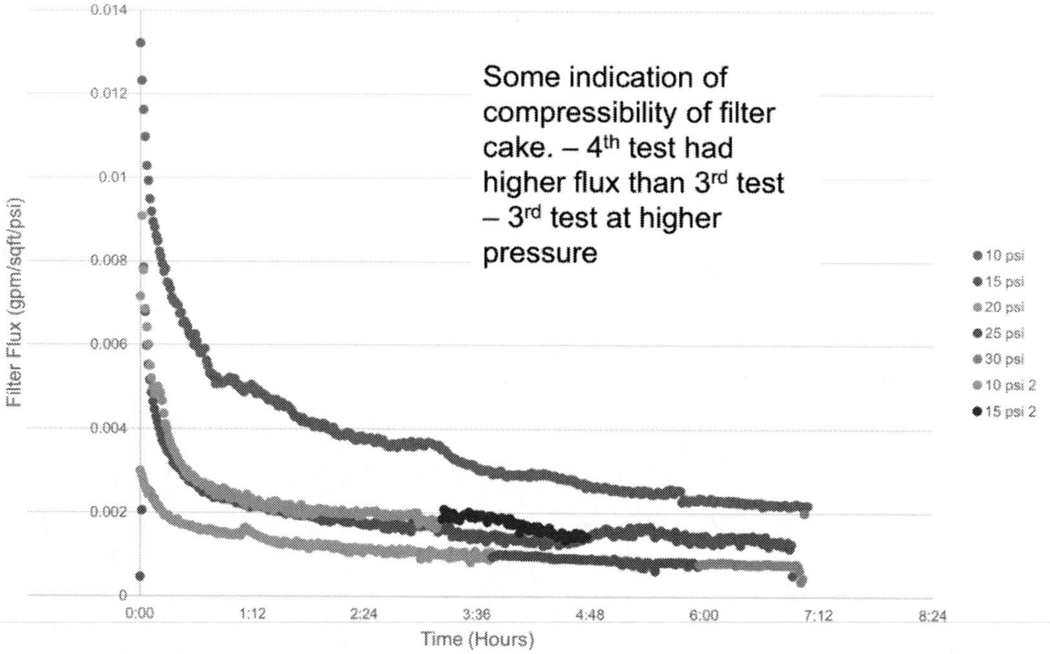

12

Proudly Operated by Battelle *Since 1965*

CUF- Dewatering Test (prototypic operation)

Compared to backpulsing, increasing TMP had limited impact on flux in terms of magnitude and duration

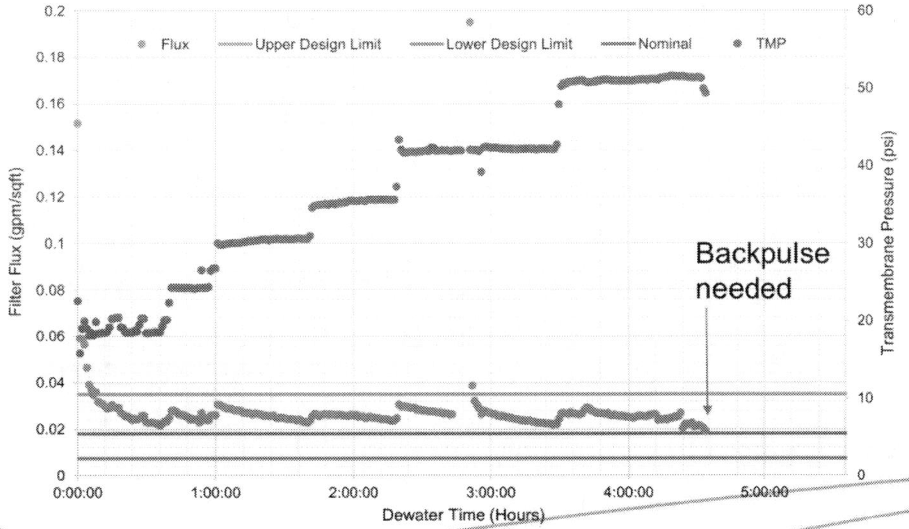

Pacific Northwest NATIONAL LABORATORY

Proudly Operated by **Battelle** *Since 1965*

13

DEF AP-105 Filtration

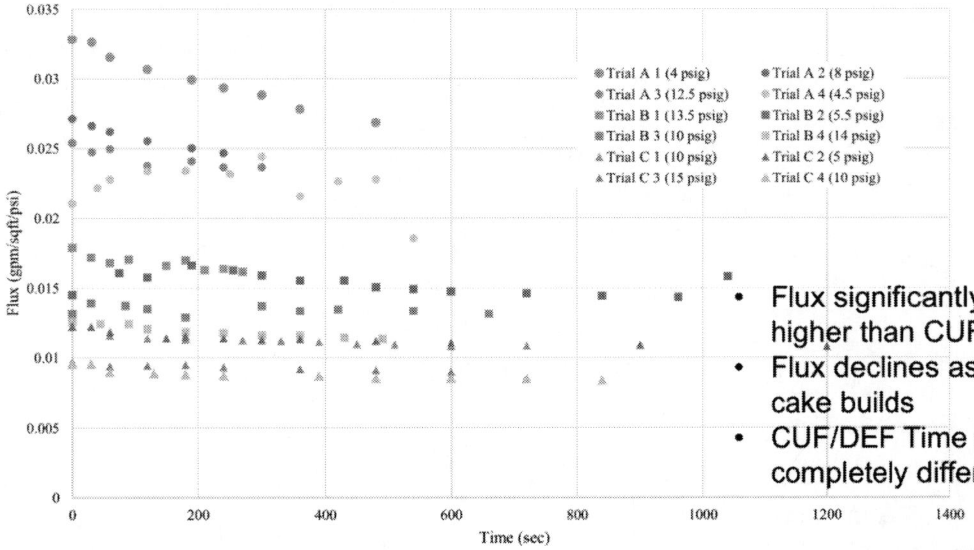

- Flux significantly higher than CUF
- Flux declines as filter cake builds
- CUF/DEF Time scales completely different

Pacific Northwest NATIONAL LABORATORY

Proudly Operated by **Battelle** *Since 1965*

14

DEF Filtration of AP-105 with varying Pressure

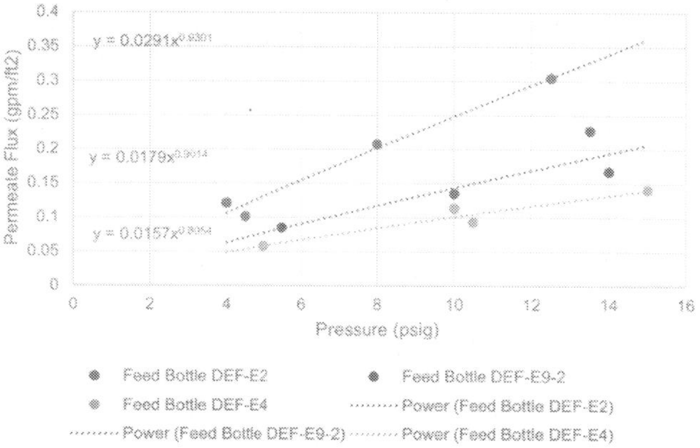

$\delta=0$ incompressible
$0<\delta<1$ moderately compressible
$\delta>1$ highly compressible

Solids are compressible

$J \propto P^{(1-\delta)}$

J- permeate flux (gpm/ft^2)
P-pressure (psig)

Pacific Northwest
NATIONAL LABORATORY
Proudly Operated by Battelle *Since 1965*

15

DEF Filtration of AP-105 at Constant Pressure

K- cake permeability (m^2)
R_m-media resistance (1/m)
$C_L - \phi_s/(\epsilon_s-\phi_s)$
ϕ_s- solid vol fraction in slurry
ϵ_s- solid vol fraction in cake

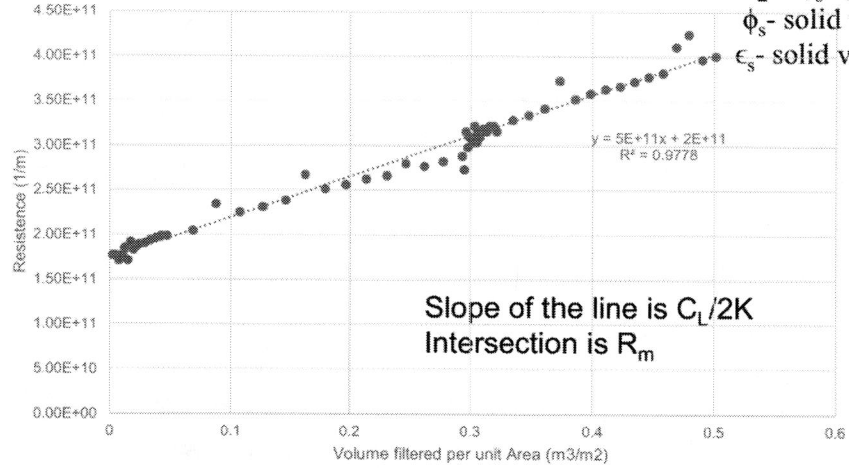

Resistance of filter cake is NOT negligible

Pacific Northwest
NATIONAL LABORATORY
Proudly Operated by Battelle *Since 1965*

16

DEF Filtration of AP-105

Observed cake thickness ~0.5mm =0.0005m
Volume Filtered per unit Area=0.5m

C_L= L/V~0.001 and K~2x10^{-15} m^2

Ease of Filtration	Permeability (m^2)
Very Easy	10^{-12}
Moderate	10^{-14}
Difficult (AP-105 Solids)	10^{-15}
Very Difficult	10^{-16}

AP-105 feed is supernatant sampled from tank farm with "no apparent" solids.

Small amount of solids present have low permeability and are considered "difficult" to filter

Pacific Northwest NATIONAL LABORATORY

Proudly Operated by Battelle *Since 1965*

Summary of Simulants Available for DEF comparison

LAWPS Simulant	Description	Density (g/mL)	Avg. CUF Long-Term Flux (gpm/ft²)	DEF Flux (gpm/ft²)	Ratio
1	5.6 M Na with ~0.5 wt% sodium oxalate solids	1.21	0.04	0.1	2.5
2	5.6 M Na with 0.1 wt% iron oxyhydroxide	1.21	0.03	0.04	1.3
3	Not used				
4	8 M Na with ~0.25 wt% precipitated solids	1.35	0.03	0.07	2.3
5	4 M Na with ~0.1 wt% boehmite solids	1.18	0.1	0.2	2.0
6	6 M Na with ~1 wt% precipitated sodium phosphate solids	1.28	0.08	0.05	0.7

Comparison is difficult (different time scales, compressible cake, growing filter cake on DEF, and filter fouling on CUF. (Flux ratio for AP-105 is ~6.)

Comparison of DEF to CUF Flux for Simulants

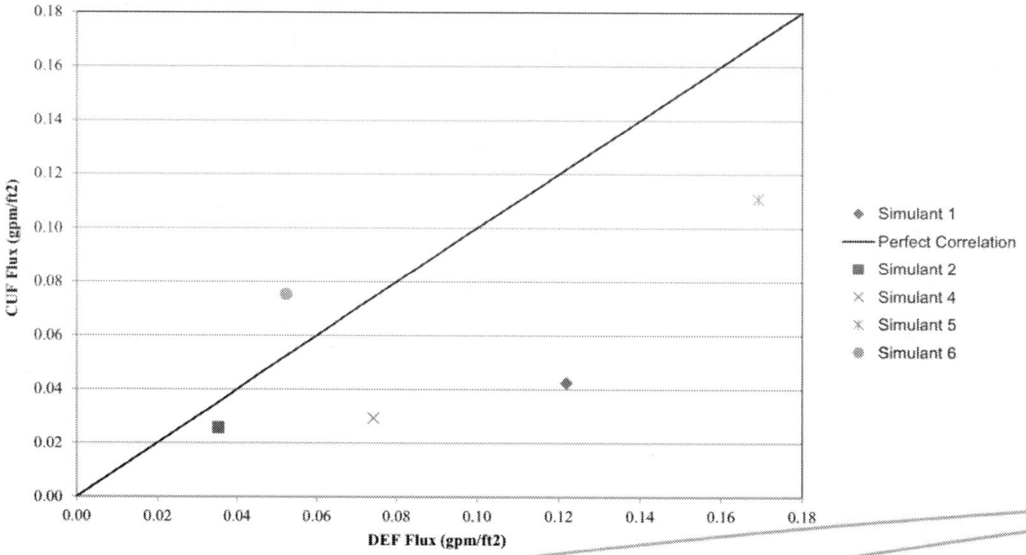

Conclusions

- Nature of the data make comparisons between DEF and CFF difficult
  - Limited correlation between DEF and CFF flux is due to fundamental differences
    - CFF method minimizes cake build up, and therefore cake resistance, allowing for more filter fouling
    - Different time scales of testing
- DEF is not a good substitute for qualifying feeds at lab scale, especially with small volume of feed
- DEF and CFF methods are complementary
  - DEF better suited for collecting and analyzing solids
  - DEF provides information on cake properties
  - CFF is prototypic of plant operations and provides better flux data and can better demonstrate backpulsing and filter cleaning methods
- AP-105 Filter cake is compressible. Demonstrated with both CFF and DEF.
- AP-105 Filter cake permeability measured. Demonstrated with DEF.
- Increasing TMP was shown not to be effective for long. Demonstrated with CUF.
- Bench scale testing shows filter cleaning will need to be done every 7 to 14 hours. Demonstrated with CUF.

BACK UP Information

Pacific Northwest
NATIONAL LABORATORY
Proudly Operated by Battelle *Since 1965*

21

Solids Residue - DEF

Pacific Northwest
NATIONAL LABORATORY
Proudly Operated by Battelle *Since 1965*

22

Solids characterization – top 10 components

(Analyte)	ppm
Na	199,500
C	92,000
Al	35,550
Cr	33,700
Fe	18,000
Mg	9,645
Ni	6,150
B	4,445
K	4,220
Pb	2,235

Na:C ratio in line with sodium oxalate

Pending additional microscope analysis

Preliminary data, still pending review

23

Pacific Northwest
NATIONAL LABORATORY

Proudly Operated by Battelle *Since 1965*

Savannah River National Laboratory
OPERATED BY SAVANNAH RIVER NUCLEAR SOLUTIONS

We put science to work.™
SRNL-STI-2017-00679

Investigation of High-Level Waste Glass Melting Using X-Ray Computed Tomography

Alexander S. Choi, Donald H. Miller, David M. Immel, and Frank G. Smith

2017 AIChE Annual Meeting, Minneapolis, MN
October 30 – November 3, 2017

Savannah River NUCLEAR SOLUTIONS

High-Level Waste (HLW) Processing at Savannah River Site (SRS)

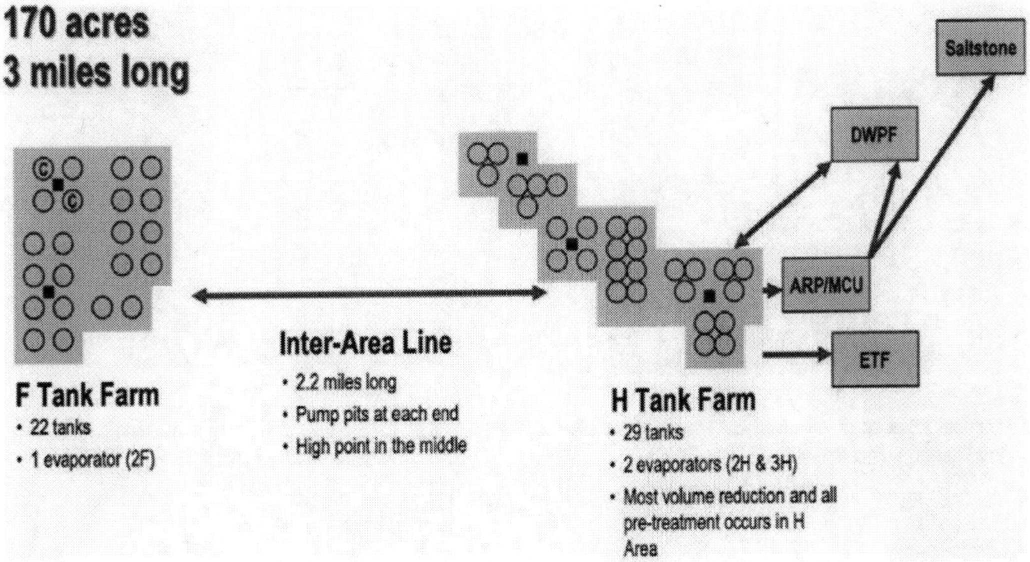

Defense Waste Processing Facility (DWPF)

- Components of HLW Glasses:
  - *Waste – Sludge (Fe, Al, Na, Mn, U, Tc, Am) and Salts (Cs, Sr)*
  - *Glass Forming Frits (Si, B, Li, Na)*
- Goals:
  - *Reduce life-cycle cost by maximizing waste throughput.*
  - *Reduce canister production by maximizing waste loading.*
  - *Optimize frit composition for maximum melt rate.*
- Current Strategy:
  - *Determine candidate frits using property/composition models.*
  - *Bench-scale melt rate testing of candidate frits to select one that yields highest melt flux – expensive and time-consuming process.*
  - *Semi-empirical modeling of melting rate.*

Melt Rate Furnace (MRF)

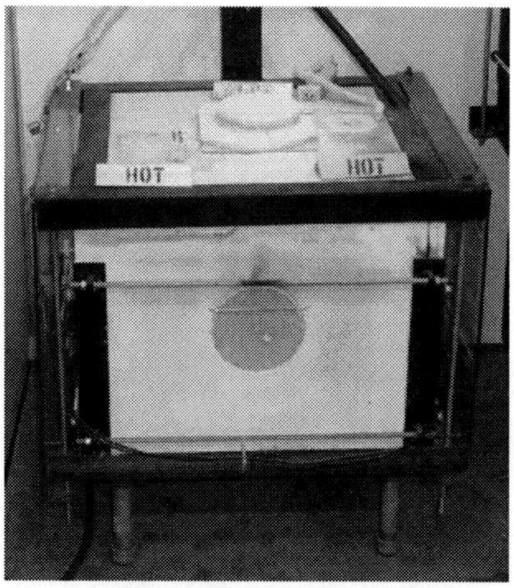

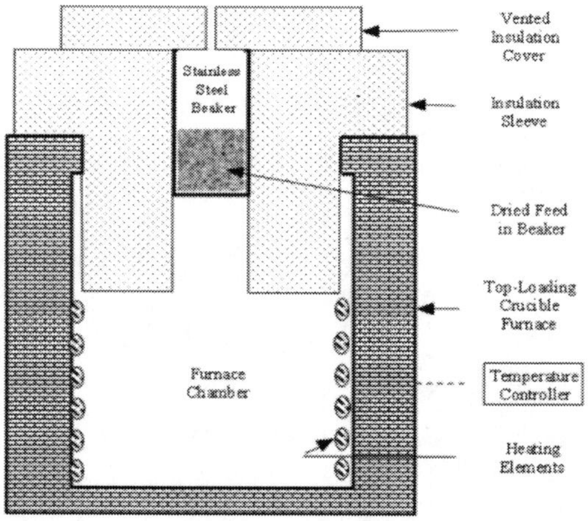

- Lindberg/Blue M crucible furnace (Model CF56822C) by Thermo Fisher Scientific.
- 19 cm diameter inner chamber with bottom 20.3 cm heated.
- 5 cm and 10 cm diameter stainless steel beakers
- Batch size = dried waste simulant/frit blend to produce ~50 g of glass when fully melted.
- One-dimensional vertical heating in the beaker to simulate the cold cap melting.
- Batch heating time = 20 - 22 min.

Visual Determination of Melting Rate

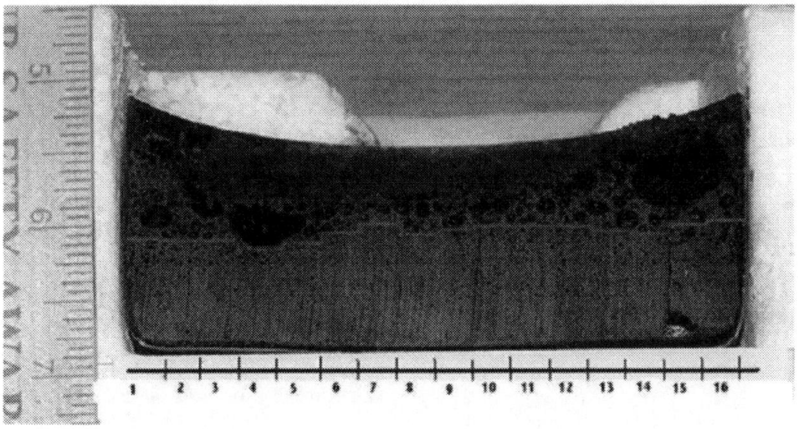

- Linear Melt Rate (LMR) is determined as the average of measured glass heights in each of the 16 horizontal segments divided by the time each sample was fired.
- LMR was useful in identifying melting accelerants and ranking relative melting rates of candidate frits.
- However, LMR becomes increasingly subjective as glass height (red line) becomes more difficult for feeds with increasing concentration of gas generators (i.e., in the presence of more numerous bubbles or voids).
- Bubble distribution and thus glass height of each cross-section will not remain the same when rotated around 360° along the center axis.

X-Ray Scan

- 5 cm-diameter beakers containing heat-treated samples were X-ray scanned using a lens-coupled digital radiography system consisting of:
  - GE/Seifert 420 Isovolt X-ray generator
  - Hamamatsu CsI scintillation screen
  - Roper Scientific Quantix CCD camera fitted with a Nikon Nikkor 105 mm lens
- A four-axis motion control system was used for sample positioning.
- Control and processing software was written using the Labview programming language.
- For each sample, 360 digital radiographs were taken at one degree angular increments.
- X-ray images were taken along a cross-section of the sample through the center axis perpendicular to the source.
- The potential at the X-ray source was set at 300 kV.
- A 1.25 mm stainless steel filter was used at the source to harden the source beam and a 0.13 mm lead filter was used at the scintillator to attenuate X-ray scatter.
- Feldkamp cone-beam reconstruction was performed on the sample data using software developed at Lawrence Livermore National Laboratory to produce 3-D volumetric data at a resolution of 0.0033 mm^3 representing a 0.15 mm sided voxel.

Computed Tomography (CT)

- A segmentation approach was used to build 3-D volumetric data of heat-treated samples at a resolution of 0.0033 mm^3 representing a 0.15 mm-sided voxel.
- Radiographic attenuation in a homogeneous material follows Lambert-Beer's Law:

$$I = I_0 e^{-\mu L} \quad (1)$$

where I_0 = intensity of the incident X-ray beam
I = intensity of the attenuated X-ray beam
μ = attenuation factor of the material in mm^{-1}
L = material dimension or length of a voxel side in mm.

- The dimensionless quantity μL = CT density (ρ_{CT}), which is correlated to material density.

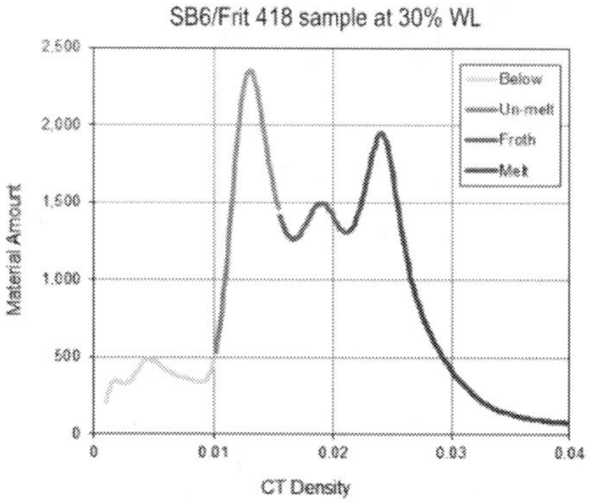

3-D volumetric data for typical samples shows four morphologically-distinct layers:

- Melt: Fully melted glass with little or no gas bubbles.
- Froth: Melted material interspersed among gas bubbles.
- Unmelt: Feed material that may have begun to soften but maintains its original granular structure.
- Below: Material of very low but non-zero density indicating gas-filled voids or radiographic artifacts.

Threshold CT Densities

- Threshold boundaries between adjacent layers were determined by differentiating and smoothing the histogram point-by-point and finding significant negative-to-positive zero crossings of the 3rd derivative of the histogram.

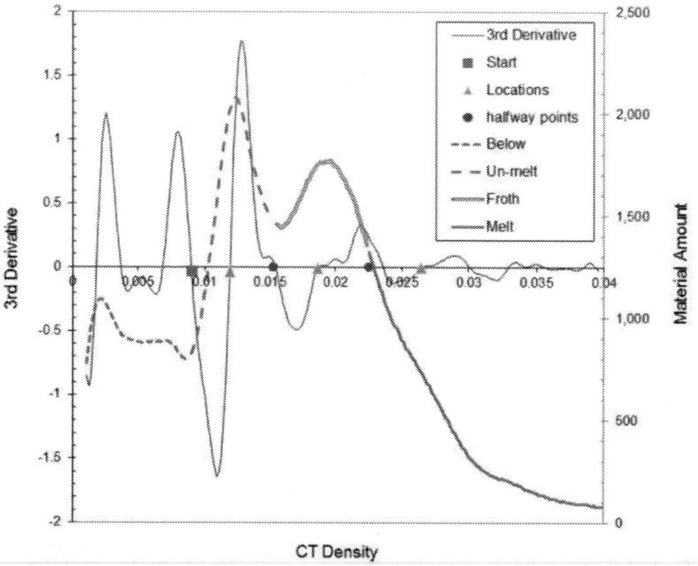

Compositions of HLW Simulants and Frit 418

Simulant	SMEM-1	SMEM-2	SMEM-3	SMEM-4	SB6
Elements (>0.1 wt%)	wt% calcined	wt% calcined	wt% calcined	wt% calcined	wt% calcined
Fe	32.00	31.60	11.15	12.30	18.90
Al	8.99	9.14	23.15	23.75	15.90
Mn (IV)	3.03	0.52	1.54	0.30	5.92
Mn (II)	1.01	0.18	1.96	0.40	
Ca	3.88	2.05	3.27	1.81	0.95
Mg	0.38	2.64	0.31	2.50	0.45
Ni	0.23	2.61	0.20	2.28	2.54
Cr	0.02	0.30	0.02	0.27	-
Cu	0.07	0.06	0.05	0.06	0.18
Si	1.50	1.51	1.42	1.32	0.14
Na	13.80	14.80	16.80	15.90	15.40
K	<0.10	<0.10	0.62	<0.10	0.07
Ce	0.11	0.11	0.10	0.10	<0.01
Ba	0.08	0.10	0.06	0.06	<0.01
Pb	0.11	0.08	0.09	0.09	<0.01
Zr	0.06	0.12	0.07	0.06	0.22
S	0.31	0.31	0.37	0.41	0.37
Total	65.58	66.13	61.18	61.61	61.04
Anions	mg/kg slurry	mg/kg slurry	mg/kg slurry	mg/kg slurry	mg/kg slurry
Cl	125	130	1,125	133	701
NO_3	22,100	27,200	22,700	25,400	45,600
SO_4	204	<100	772	<100	931
COOH	51,400	48,900	51,100	47,300	89,000
C_2O_4	<100	<100	<100	<100	399

Target Composition of Frit 418

Oxide	wt %
B_2O_3	8
Li_2O	8
Na_2O	8
SiO_2	76
Total	100

X-Ray Images of Bounding Sludge/Frit 418 Feeds at 36% Waste Loading (WL)

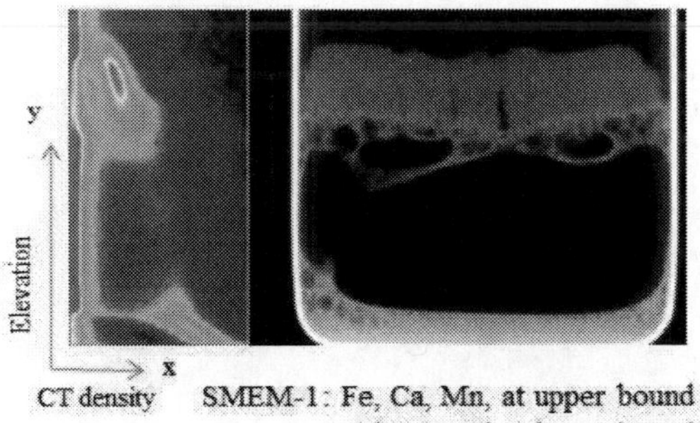

SMEM-1: Fe, Ca, Mn, at upper bound
Al, Mg, Ni at lower bound

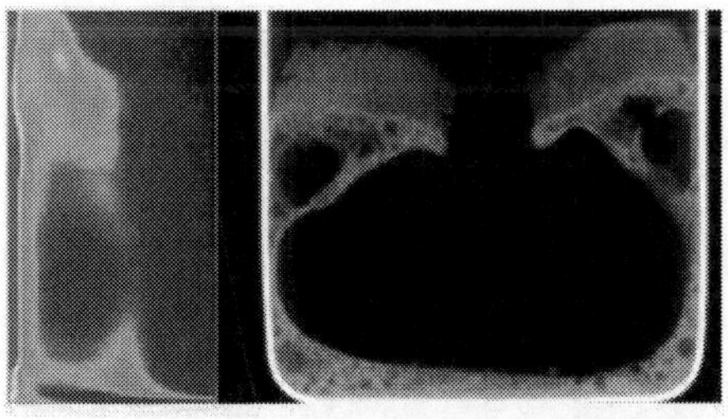

SMEM-2: Fe, Mg, Ni at upper bound
Al, Ca, Mn at lower bound

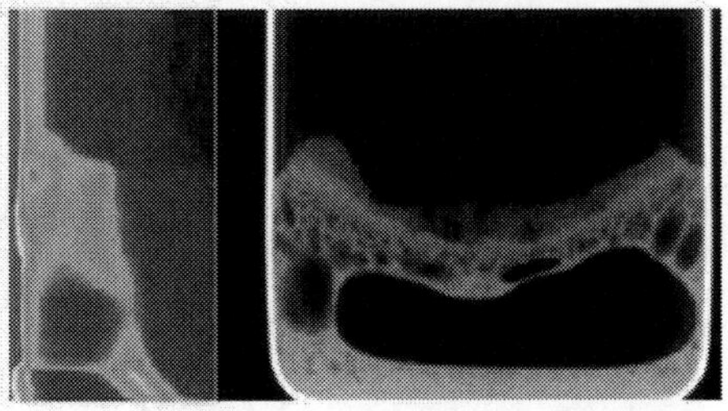

SMEM-3: Al, Ca, Mn at upper bound
Fe, Mg, Ni at lower bound

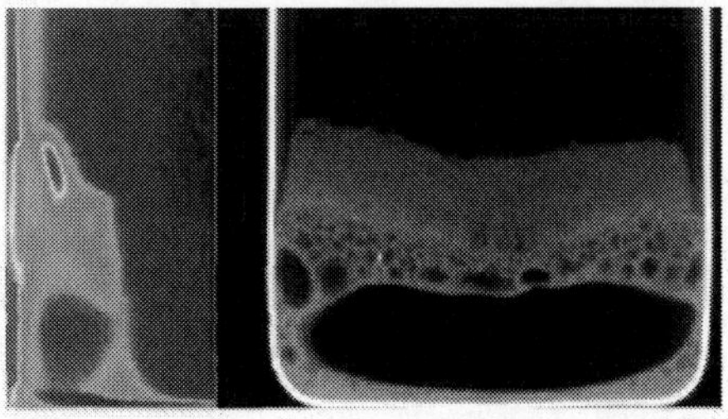

SMEM-4: Al, Mg, Ni at upper bound
Fe, Ca, Mn at lower bound

Histograms of Bounding Sludge/Frit 418 Feeds

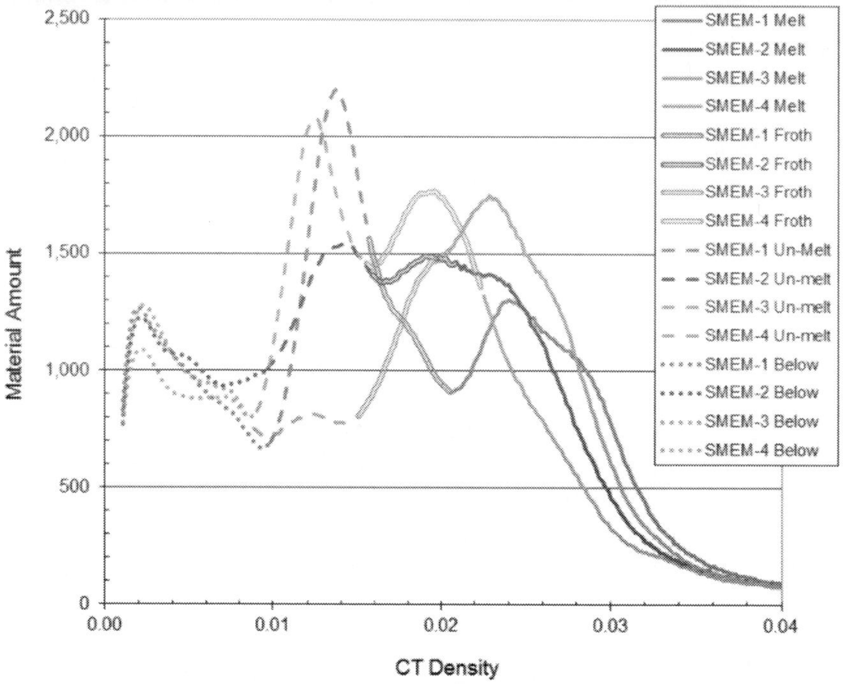

Material Amount and Density of Heat-Treated Bounding-Sludge/Frit 418 Feeds

Sample ID	Material Amount					Material Density (mm^{-3})				
	Total	Un-melt	Froth	Melt	Melt + Froth	Overall	Un-melt	Froth	Melt	Melt + Froth
SMEM-3	161,430	27,064	40,805	93,562	134,366	5.784	3.417	5.347	7.572	6.722
SMEM-2	165,923	47,278	46,189	72,456	118,645	5.414	3.716	5.462	7.656	6.620
SMEM-1	174,012	57,499	36,410	80,103	116,513	5.453	3.866	5.314	7.863	6.838
SMEM-4	160,654	55,857	58,867	45,930	104,797	4.967	3.551	5.423	7.978	6.308
std dev/ mean	3.7%	29.8%	21.3%	27.5%	10.3%	6.2%	5.4%	1.3%	2.4%	3.4%

Projected Composition of SMEM Glasses and Calculated Properties

Oxides:	SMEM-1 (mol%)	SMEM-2 (mol%)	SMEM-3 (mol%)	SMEM-4 (mol%)
Fe_2O_3	6.88	6.68	2.34	2.52
Al_2O_3	4.00	4.00	10.05	10.07
MnO	1.76	0.30	1.49	0.29
CaO	2.32	1.21	1.91	1.03
MgO	0.37	2.57	0.30	2.36
NiO	0.09	1.05	0.08	0.89
Cr_2O_3	0.01	0.07	0.00	0.06
CuO	0.02	0.02	0.02	0.02
SiO_2	54.80	54.39	53.30	52.84
Na_2O	13.28	13.41	14.22	13.71
K_2O	-	-	0.18	-
Ce_2O_3	0.01	0.01	0.01	0.01
B_2O_3	4.86	4.83	4.74	4.70
Li_2O	11.33	11.25	11.03	10.96
BaO	0.07	0.08	0.05	0.05
PbO	0.01	0.01	0.01	0.01
ZrO_2	0.02	0.03	0.02	0.01
SO_3	0.14	0.08	0.22	0.44
Total	99.97	99.99	99.97	99.97
net alkali ion field strength* (1/Å^2)	69.997	70.203	67.341	65.254
non alkali ion field strength (1/Å^2)	15.807	28.337	13.168	25.328
total modifier field strength (1/Å^2)	85.804	98.540	80.510	90.582
calculated viscosity at 1,150 °C (Pa·sec)	3.445	3.338	13.086	13.879
calculated liquidus, T_l (°C)	937.6	1,068.3	822.0	946.9

Factors Affecting HLW Glass Melting Rate

- Total field strength (*FS*) of modifier ions is calculated as the weighted sum of individual modifier field strengths:

$$FS = \sum_{k=1}^{n} C_k \left(Z_k / a_k^2\right) \qquad (2)$$

where C_k = concentration of modifier *k* (as oxide) in mole fraction
Z_k = valence charge of modifier *k*
a_k = effective ionic radius of modifier *k* in Å

The lower the *FS*, the less rigid the network structure (hence, lower melt density), allowing more large radionuclides into the network.

- The higher the rate of gas evolution from salt decomposition, the lower the rate of heat transfer.

As the nitrate decomposes over a wider temperature range than the formate, it has a greater adverse impact on the melting rate.

- The higher the concentration of *Al*, the more modifier ions are tied up with AlO_4^- tetrahedra, making the network structure more rigid and the melt viscosity higher.

- However, the higher the melt viscosity, the more sluggish the melt in closing the channels through which evolving gases escape.

X-Ray Images of Heat-Treated SB6/Frit 418 Feeds at Varying WL

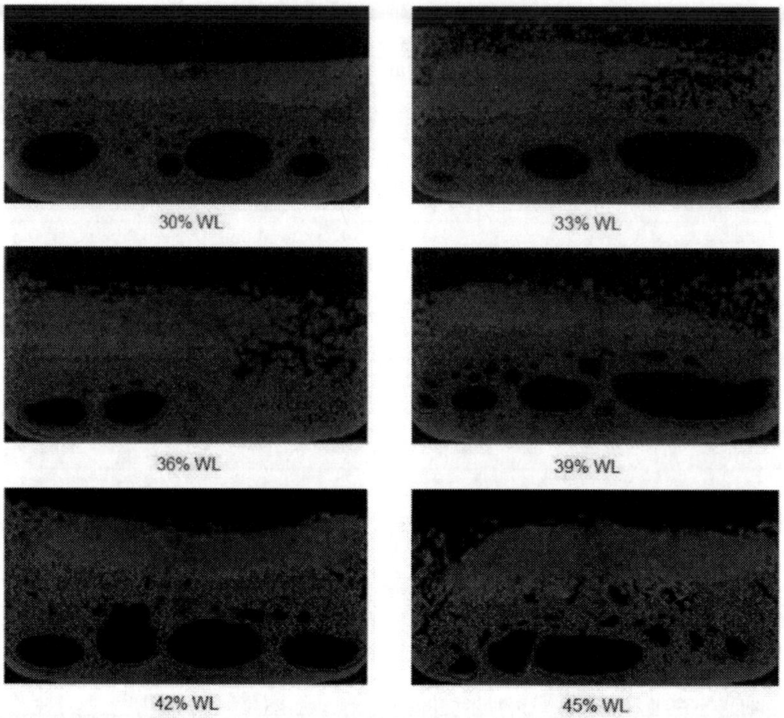

Histograms of SB6/Frit 418 Feeds at Varying WL

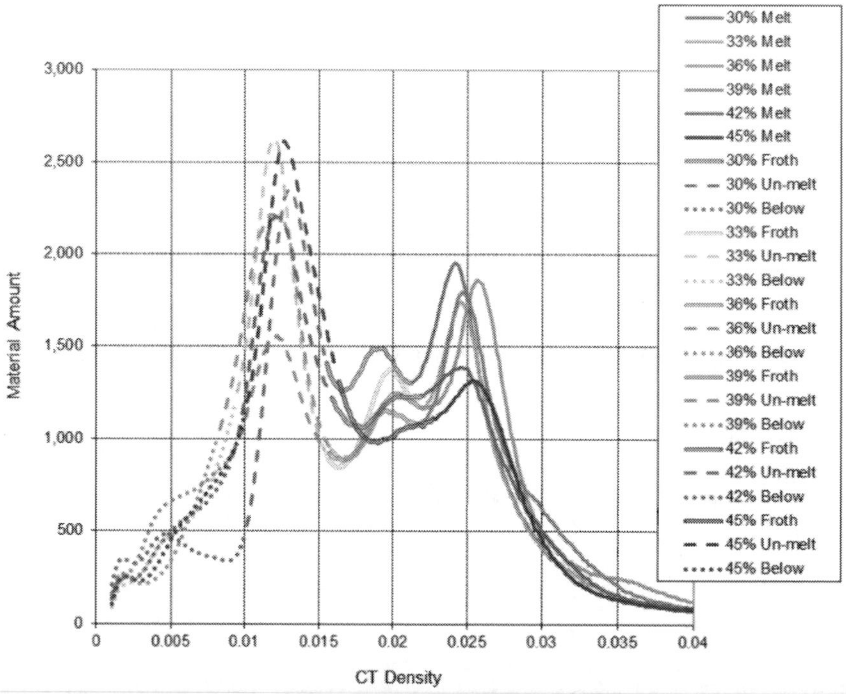

Material Amount and Density of Heat-Treated SB6/Frit 418 Feeds

Waste Loading (wt%)	Material Amount					Material Density (mm^{-3})				
	Total	Un-melt	Froth	Melt	Melt + Froth	Overall	Un-melt	Froth	Melt	Melt + Froth
30	180,710	55,996	49,417	75,297	124,714	5.399	3.854	5.451	7.623	6.584
33	176,536	66,460	43,512	66,563	110,076	5.070	3.595	5.589	7.789	6.740
36	195,145	75,999	43,380	75,766	119,146	5.015	3.474	5.690	8.048	6.993
39	166,307	50,761	38,222	77,324	115,546	5.396	3.582	5.557	7.910	6.938
42	177,440	68,310	45,327	63,803	109,130	5.106	3.662	5.624	7.935	6.778
45	174,004	74,998	42,602	56,405	99,007	5.150	3.868	5.760	8.054	6.876
std dev/ mean	3.2%	25.6%	61.0%	39.9%	32.1%	3.2%	4.3%	1.9%	2.1%	2.2%

Relative Melting Rates vs. Waste Loading of SB6/Frit 418 Feeds

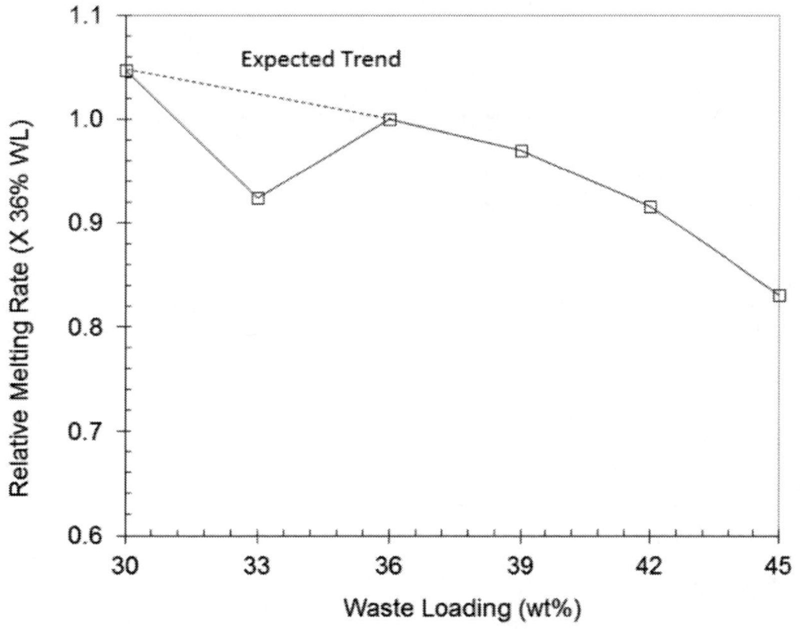

Reproducibility of Data

- The reproducibility of CT results was checked using duplicate samples of:
  - Four different simulant compositions.
  - Each simulant spiked with high and low concentration of gas generators (or salts).
- The concentration of salts showed no appreciable effect on the scatter of data.
  - The standard deviation ranged from 3.9 to 13.7% of their respective means for feeds with higher concentrations of salts and from 0.02 to 15.7% for feeds with lower concentrations of salts.
- Due to small sample size, the root-mean squared error (RMSE) was used as a generalized standard deviation encompassing all four feeds:

Relative RMSE of All Samples

Unmelt	Froth	Melt	Melt+Froth
7.2%	4.5%	8.2%	1.7%

Relative RMSE of Two Samples Re-Scanned One-Year Later

Unmelt	Froth	Melt	Melt+Froth
1.7%	1.1%	1.1%	1.0%

Conclusions

- X-ray CT is capable of providing:
  - Visual images of batch-to-glass conversion, including gas evolution and resulting bed expansion.
  - A quantitative measure of relative melting rates of various waste/frit blends in a consistent and objective manner.
  - A framework for optimizing frit design to accommodate ever-changing waste batch chemistry.
- X-ray images of heat-treated samples revealed the presence of four morphologically distinct layers: Melt, Froth, Unmelt and Below, as defined earlier.
- Quantifying the amount of material in each layer by CT and combining the material amounts in the Melt and Froth layers would give a viable measure of relative melting rate rankings that is representative of continuously-fed melters.
- The HLW glass melting rate is at least affected by:
  - Total field strength of modifier ions
  - Rate of calcine gas evolution
  - Concentration of Al
  - Melt viscosity

Acknowledgements

- Funding was provided by:
  - U.S. Department of Energy EM-31 Technology Development & Deployment (TDD) Program
  - U.S. Department of Energy Office of River Protection (DOE-ORP) at Hanford, WA
- Dr. Pavel Hrma of Pacific Northwest National Laboratory (PNNL) and Dr. Albert Kruger of DOE-ORP for technical review of the paper.
- This work has been published in *International Journal of Applied Glass Science (2017)*.

503054 A Multi-Phase Continuum Approach to Modeling the Performance of a Fluidized Bed Nuclear Reactor

Abstract:

The Generation IV International Forum has identified six designs to be considered for the future nuclear power systems to be built and operated in 2030 and forward. Of the six designs, two are gas cooled, two are liquid metal cooled and one is a molten salt reactor. Research in the area of fluidized bed systems is of importance to several of the potential Generation IV reactor designs. The conceptual design of the fluidized bed nuclear reactor proposed by Kloosterman, Golovko, van Dam & van der Hagen (2001) is a simple design consisting of a vertical graphite-walled tube with graphite bottom and top reflectors partially filled with TRISO-coated fuel particles made of UO_2 fuel kernel surrounded by three layers of pyrolytic carbon and silicium carbide modeled as one moderator shell. Helium coolant flows from bottom to top through the reactor to cool the fuel particles and when its flowrate reaches a critical velocity, the particle bed expands. As a result, more neutrons are allowed to escape to the graphite walls where they are moderated and the reactivity of the core increases. However, if the bed expands beyond a critical core height, neutrons start to leak away and reactivity decreases. The expanded core height is strongly dependent on the particle size and density distributions in the core. Moreover, instabilities in the form of bubbles and particle clusters, which can form due to size and/or density distributions, can significantly affect reactivity and core neutronics requiring further investigations.

Fluidized beds are currently used extensively in the petrochemical and pharmaceutical industries and there is extensive chemical engineering modeling work on the steady state and time-dependent hydrodynamics of gas- and liquid-fluidized beds with one particle species (see Jackson, 2000). However, models incorporating particle size and density distributions have been examined to a much lesser extent. This work considers a multiphase approach which makes use of volume-averaged mass and momentum balances for the fluid and particle phases independently and generalizes the monocomponent equations proposed by Anderson & Jackson (1967). The model consists of a multi-solid mixture of fluidized particles consisting of N particle species that mimics the particle size and/or density distribution of the TRISO fuel particles. The linear stability of the fluidized reactor core will be examined using hydrodynamic stability theory and a multi-fluid continuum approach to determine if and when the fluidized reactor core becomes hydrodynamically unstable. A reactor model, consisting of the proposed multi-phase continuum model along with a thermal hydraulics model and a neutron kinetics model, can be combined to investigate the dynamic stability of the reactor and the relationship between the expanded reactor core height, core neutronics, reactivity and coolant mass flowrate.

494842 Large-Scale Atomistic Simulations of Low-Energy Helium Implantation into Tungsten Single Crystals

Abstract:

Large-scale molecular dynamics simulations of post-implantation helium behavior in plasma-facing tungsten single crystals reveal orientation-dependent depth profiles, surface evolution patterns, and other crystallographic and diffusion-related characteristics of helium behavior in tungsten during the first microsecond. The flux of implanted helium atoms studied, $\Gamma = 4 \times 10^{25}$ m^{-2} s^{-1}, is only one order of magnitude larger than that expected ITER, the experimental fusion reactor currently being constructed in France. With simulation times on the order of one microsecond, these results serve to discover of the mechanisms involved in surface evolution as well as to serve as benchmarks for coarse-grained simulations such as kinetic Monte Carlo and continuum-scale drift–reaction–diffusion cluster dynamics simulations. The findings of our large-scale simulations are significant due to diminished finite-size effects and the longer times reached (corresponding to higher fluences). Specifically, our findings are drastically different from findings published previously in the literature for (0 0 1) surfaces under a helium flux of $\Gamma \sim 10^{28}$ m^{-2} s^{-1}, which is typical of smaller size and shorter time atomistic simulations. In particular, this study highlights the atomic-scale materials processes relevant to helium entrapment and transport in metals, which have implications not only for nuclear fusion–relevant processes, but also helium-induced embrittlement in irradiated materials such as hospital equipment and fission reactor materials.

498065 Development of Electrochemical Process Models for Treatment of Used Nuclear Fuel

Abstract:

Chemical process models that simulate the unit operations within an electrochemical metallic used-fuel treatment facility are being developed to inform the design and implementation of process monitoring systems for material accountancy. The Argonne Model for Pyrochemical Recycling (AMPYRE) calculates the mass balance of a complete electrochemical reprocessing facility. The code consists of a series of interdependent unit operations models. Calculations enable prediction of the batch-wise evolution of salt and product compositions. The code follows all of the fuel components (U, TRU, rare earth and active metal fission products), plus Li, K, from the salt and Na from the fuel bond. The anode and cathode subunits use thermodynamic parameter values obtained from the open literature or supplied by the user. All major material movements are captured, including salt recycle from cathode processing and waste treatment. As a comprehensive representation, the code can be used to design flowsheets or to study operational effects on output. The Dynamic Electrorefiner (DyER) code simulates the major components of an electrorefiner: the anode, U cathode, U/TRU cathode, and molten salt electrolyte. Built in Mathworks MATLAB®, it is a time-dependent code that uses an ODE solver to predict the compositional evolution of these various components. It can be run as either standalone software, or interfaced with a facility or sensor model to provide a more rigorous representation of electrorefiner operations. DyER follows the same species as the AMPYRE code. As a fully dynamic model, DyER can be used to explore transient conditions in the electrorefiner and to predict the outcomes arising from time-variant or off-normal inputs or conditions. These codes predict the evolution of the distribution of uranium and the transuranics (TRU), as well as key fission products, in a facility when direct measurement is not practical during process operations, or when a fully integrated set of processes is not available. The calculation of the in-process inventory serves as part of an integrated approach to safeguards by informing the design and selection of measurement techniques, sensor arrangement, and the development of specialized analysis tools.

The submitted manuscript has been created by the UChicago Argonne, LLC as Operator of Argonne National Laboratory (â€œArgonneâ€) under Contract No. DE-AC02-06CH11357 with the U.S. Department of Energy. The U.S. Government retains for itself, and others acting on its behalf, a paid-up, nonexclusive, irrevocable worldwide license in said article to reproduce, prepare derivative works, distribute copies to the public, and perform publicly and display publicly, by or on behalf of the Government.

503053 Managing Spent Fuel Resources – Closing the Loop

Abstract:

Wide spread deployment of nuclear power plants is curtailed due to three critical issues associated with the nuclear fission technology – safety (plant accidents and nuclear waste), security (facilities and material) and safeguards (proliferation of nuclear weapons material). These three inherent technology specific challenges coupled with energy economics and policy continues to plague nuclear power to realize its full potential. The development in electrometallurgical processing techniques and Gen IV reactors are some of the key enabling technological advancements that could shape the the future for nuclear fuel management. The elements of nuclear – chemical fundamentals is inherent to this new developments.

The global spent nuclear fuel (SNF) inventory will reach 400,000 tonnes by 2025. The SNF inventory in US reaching 70,000 MTU peaking spent pool storage capacity at several sites within US. In early 1980s the nuclear power plants began to move spent fuel assemblies to on-site dry cask storage to address this crisis. Many nuclear power countries, France, UK, Russia and Japan among others, use reprocessing and recycling as a strategy for management of SNF. China is rapidly expanding it reprocessing capacity to close the Chinese nuclear fuel cycle and to manage SNF inventory. India due to its low uranium resources and an interest in thorium fuel cycle has built in reprocessing as part of its fuel resource management strategy. South Korea, currently does not reprocess, has invested significantly in the development of the pyroprocessing technology to manage the rapidly growing SNF inventories (~700 tons per year) in Korea. The United State pioneers in the development reprocessing technologies put a moratorium on reprocessing of commercial SNF in late 1970s with the overarching goal to halt the spread of fuel reprocessing technologies worldwide. Although, the global reprocessing capacity is about 4500 tons per year and continues to grow as more countries add nuclear power to their energy mix.

Besides nonproliferation concerns there are several other factors that contribute to the limited viability of reprocessing as a comprehensive SNF management option today. The reprocessing technology, PUREX, is hydrometallurgical process that separates uranium and plutonium. Recycling reprocessed uranium (RepU) directly into the LWR is challenging due to the presence of U-232 (radiotoxicity) and U-236 (high neutron absorption cross-section). Presence of these two isotopes in RepU require higher enrichment and increased fuel handling cost. Hence currently RepU is typically stored and not recycled. The plutonium from the PUREX has to be used rather quickly to reduce risks of proliferation and to limit the production of Am-241 (radiotoxicity). PUREX process does not produce a separated stream of minor actinides leading to a waste stream with mixed long- and short lived isotopes a distinct negative from waste management prospective.

486946 Modeling of Nano-Fuzz Formation in Helium-Ion-Irradiated Tungsten

Abstract:

Due to its low hydrogen solubility, low sputtering yield, high melting point, and high thermal conductivity, tungsten (W) is considered as a suitable plasma-facing material (PFM) candidate for divertor and first-wall systems, capable of tolerating the extreme conditions of high temperature and particle flux inside fusion reactors. However, experiments have shown that helium (He) from linear and tokamak plasma devices is responsible for the formation of a nanostructure with a fuzz-like morphology on the W surface after a few hours of plasma exposure. We are especially interested in fuzz formation under the operating conditions of temperature,Â He impact energy, and He flux expected for ITERâ€™s divertor, which affect the reactor performance leading to increases in the nucleation of He bubbles, retention of hydrogen isotopes, and production of high-atomic-number dust.

In this presentation, we focus on obtaining a fundamental understanding of the initial stage of fuzz formation and predicting the surface morphological evolution of helium-ion-irradiated tungsten considered as a PFM based on an atomistically-informed, continuous-domain model that we have developed. According to this model, we have conducted a systematic protocol of self-consistent dynamical simulations of the evolution of the irradiated tungsten surface morphology, focusing on W{110} surfaces, in order to identify the critical range of conditions for nanotendril formation on the surface, a precursor to fuzz-like surface growth. We examine a broad range of operating conditions, including surface temperatures from 1300 to 2300 K, He ion energies from ~10 eV to ~1 keV, and He fluxes over several orders of magnitude from 10^{16} to 10^{22} m^{-2} s^{-1}. We compare our simulation results with recent experimental measurements and find the predicted size and arrangement of nanotendrils to be in good agreement with the experimental observations. We also present the results of a sensitivity analysis of the key model parameters, such as He concentration and He nanobubble size. Future model extensions driven by comparisons of the model predictions with the experimental observations, as well as the anticipated divertor performance, will be discussed.

500857 Modeling Xenon Transport in Molten Salt Fueled Reactors

Abstract:

Modeling Xenon Transport in Molten Salt Fueled Reactors

Valmor F. de Almeida[1], Benjamin S. Collins[1], Robert K. Salko[1], and Robert Z. Taylor[2]

[1]Oak Ridge National Laboratory, Oak Ridge, TN 37831-6181

[2]University of Tennessee, Knoxville, TN

Abstract

There exists a renewed interest in molten salt nuclear reactor (MSR) technology in view of its favorable safety attributes, namely, 1) the fuel is already in the liquid state (hence no fuel melt down concerns), 2) operation near ambient pressure, 3) no fire or explosion hazard, 4) with loss of pumping power, the fuel is drained by gravity from the reactor's core into storage tanks retaining the fission products in the frozen salt. There are also economic attributes that justify the revival of the molten salt reactor development, and there are disadvantages and concerns that need to be resolved before this reactor type can be adopted. Nevertheless, our focus is in one aspect of any MSR design that must be implemented, that is, the removal of volatile fission products during the reactor's operation. This is typically accomplished by contacting the molten salt with an inert gas, *e.g.*, He, either as cover gas in the pump assembly or as entrained bubbles into the salt (or both). Noble gases are typically removed from the salt in this operation mode, in particular ^{135}Xe which is the most notable thermal neutron absorber of the volatile fission products. Ability to continuously remove volatile fission products is a key enabling feature of high-burnup MSR's and their economic advantage over existing commercial reactors.

The addition of a gas to a molten salt flow significantly complicates the modeling and simulation of the coupled mass, heat, momentum, and neutron transport in the presence of fission, transmutation and decay. We are developing modeling and simulation capability for coupled transport in MSR's and this talk describes current efforts to develop an integrated simulation tool. Our point of the departure as far as data and model are concerned is the Oak Ridge National Laboratory Molten Salt Reactor Experiment which operated in 1965-1969. We are specifically developing a model for the removal of ^{135}Xe by means of He bubbles in the LiF-BeF_2 salt mixture with soluble UF_4 fuel under reactor operation conditions. The distinct difficulties faced at this point are: the development of a non-isothermal, two-phase, volume averaged fluid flow model with volatile mass transport from the liquid phase into the gas phase, diffusion/dispersion, and generation of the fission products in the liquid phase, and coupling of speciation to neutron transport. The computational platform used is leveraging ORNL codes from the Consortium for Advanced Simulation of Light Water Reactors (CASL), a DOE Energy Innovation Hub.

This talk will also provide an outlook of the additional modeling of redox reactive fluid flows needed to account for corrosion of vessel components and deposition of metallic fission products as interfacial transport phenomena.

Vapor–liquid Phase Equilibria Diagram For Uranium Hexafluoride (UF_6) Using Simplified Temperature Dependent Interaction Potential

ALI KH. AL-MATAR[1] (aalmatar@ju.edu.jo)
HOUSAM BINOUS[2] (binoushousam@yahoo.com)

Department of Chemical Engineering

[1] University of Jordan, Amman 11942, Jordan
[2] King Fahd University of Petroleum & Minerals, Saudi Arabia

Presentation Outline

- Importance of UF6
- Reasons behind this work
- Scope and Methodology
- Anticipated Objectives
- Experimental Data for the Second Virial Coefficient vs. Temperature for UF6
- VLE data for UF6
- Results for the parameters of the L-J Potential
  (both for the TIIP and the TDIP cases)
- VLE Predictions, Vapor Pressure and Critical Properties from the GEMC simulations
- Conclusions
- Future work

Importance of UF_6

- Uranium hexafluoride (UF_6, CAS: 7783-81-5).

Strategic importance in the nuclear industry

- Precursor for the enrichment process
- Product of the volatilization of spent fuel reprocessing.

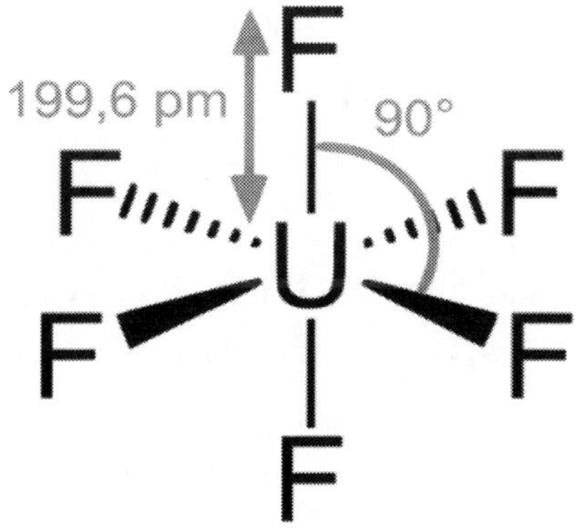

Motivation

- Data in the open and grey literature pertinent to the thermophysical properties and phase coexistence of UF_6 have discrepancies.
  - Particularly true for the vapor-liquid equilibrium properties.
  - Problem compounded for the near and supercritical region where virtually experimental data are practically nonexistent.

Research Problem

- The main proposition in this research is that introducing temperature dependent interaction parameters (TDIP) for the Lennard-Jones potential (collision diameter σ and well depth ε) instead of the current practice of using temperature independent interaction parameters (TIIP),

- This study aims at comparing phase diagram of UF_6 using TDIP and TIIP.

- Among the objectives: better prediction of phase equilibria

Scope and Methodology

- This work is focused on obtaining the vapor-liquid equilibrium (VLE) curve for uranium hexafluoride (UF_6) using TDIP as opposed to TIIP.

How?

1- The interatomic potential parameters are estimated using second virial coefficient data with simplified temperature dependence as obtained in Al-matar and coworkers previous work [1-4].

2- Subsequently, the simplified TDIP will be used to generate the VLE curve using Gibbs Ensemble Monte Carlo (GEMC) simulation.

1. Al-Matar, A.K., A.H. Tobgy, and I.A. Suleiman, The phase diagram of the Lennard-Jones fluid using temperature dependent interaction parameters. Molecular Simulation, 2008. 34(3): p. 289-294.
2. Al-Matar, A., A generating equation for mixing rules and assessment of their effect on the second virial coefficient., in Chemical Engineering2002, New Mexico State University: Las Cruces.
3. Al-Fayyd, M.A.A., *Self-diffusion coefficient of Lennard-Jones fluid using temperature dependent interaction parameters*, in *Chemical Engineering Department*2010, University of Jordan: Amman, Jordan.
4. Al-Matar, A.K., et al., *Improving Monte-Carlo and Molecular Dynamics Simulation Outcomes Using Temperature-Dependent Interaction Parameters: The Case of Pure LJ Fluid.* International Journal of Computational Methods, 2015. **12**(02): p. 1550003.

Objectives

The main objective of this research work is to obtain the VLE curve for pure UF_6.

The sub objectives include:

1- Obtain the best temperature dependent interaction parameters for the collision diameter and well depth of UF_6 using the Lennard-Jones potential based on the second virial coefficient data.

2- Generate the VLE phase diagram of pure UF_6 using temperature dependent and temperature independent interaction parameters.

3- Obtain the critical constants of UF_6 from the generated phase diagram.

4- Compare both approaches.

Literature Review: Second Virial Coefficients (SVC)

- The most recent compilation by Dymond *et al.* recommended only two data sets for UF6.
- Yet, even these two data sets present some discrepancies. They have recommended the values to be determined from equation whose coefficients were obtained by a weighted least square fit of these selected experimental values.

$$B\,(\mathrm{cm^3/mol}) = -1.6977\times10^{3} + \frac{1.8712\times10^{6}}{T\,(\mathrm{K})} - \frac{7.7035\times10^{8}}{T^{2}} + \frac{7.7828\times10^{10}}{T^{3}}$$

1. Dymond, J.H., et al., *Virial coefficients of pure gases and mixtures, Subvolume A: Virial coefficients of pure gases*2002, Berlin: Springer-Verlag.

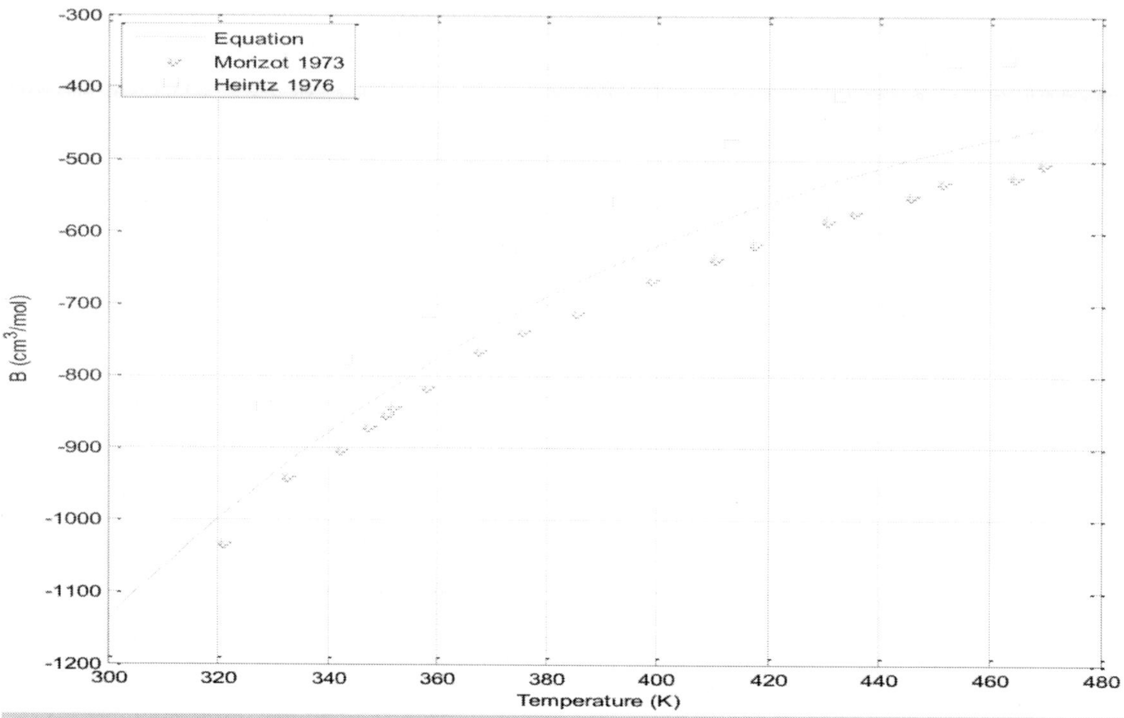

Plot of the experimental SVC from Heintz and Morizot vs. temperature and the recommended equation by Dymond *et al.*

Clearly the equation is weighted towards the data of Morizot more than those of Henitz.

1. Heintz, A., E. Meisinger, and R. Lichtenthaler, *Meßwerte des zweiten Virialkoeffizienten und Bestimmung des zwischenmolekularen Wechselwirkungspotentials von gasförmigem Uranhexafluorid.* Berichte der Bunsengesellschaft für physikalische Chemie, 1976. **80**(2): p. 163-166.

2. Morizot, P., J. Ostorero, and P. Plurien, J. Chem. Phys. , 1973. **70**: p. 1582.

Analysis of the Status of the SVC for UF_6

- The second virial coefficient data, though available, is scattered and their uncertainty is rather high. Differences between available data sets amount to 15%.
- This variation has a huge impact on the fitted results of the well depth and collision diameter.
  - The obtained rough range is 4-9 for collision diameter and 200-1000 for the well depth is indicative of the effect of this variability.
- In consequence, such variation will affect the accuracy of any subsequent calculations be it thermophysical, dynamic or spectroscopic.

Literature Review: Vapor-Liquid Equilibria

- The vapor-liquid equilibrium data available for UF_6 are rather rare and have some discrepancies.
- The phase diagram and density of saturated liquid are available. However, saturated vapor density is not available.
- Many researchers resort to using EOS to estimate the saturated vapor density. There is a need to have some experimental measurements for this property.
- The VLE phase diagram is extrapolated near the critical point and further investigation on the shape of the curve is needed.

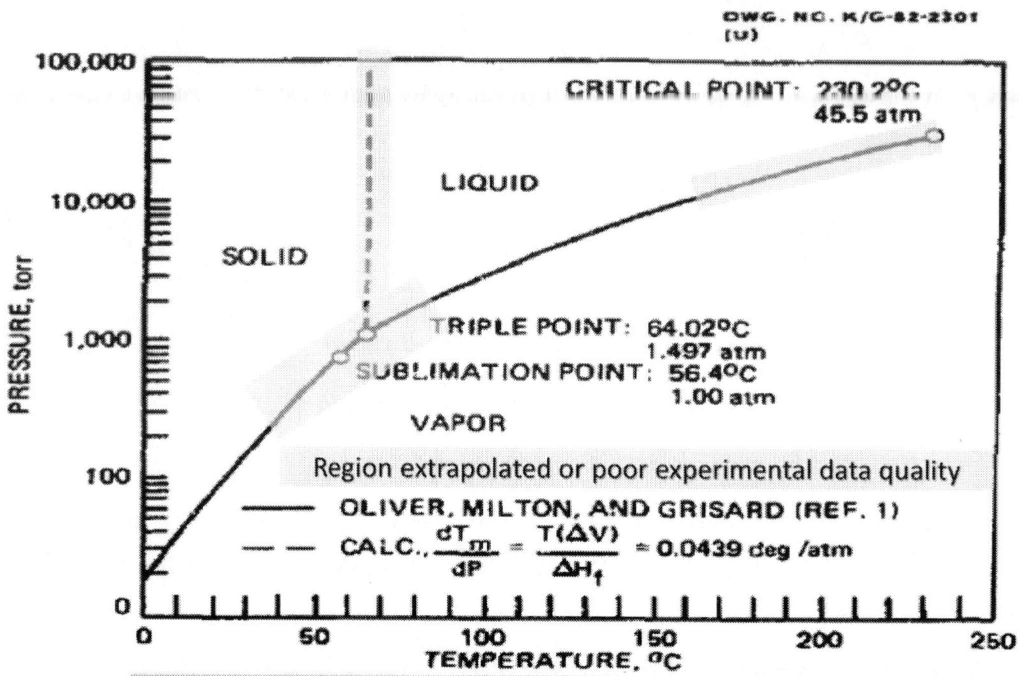

Phase diagram of UF_6.

Barber, E., *The physical and chemical properties of uranium hexafluoride.* Uranium Hexafluoride-Safe Handling, Processing, and Transporting, 1988: p. 1.

Potential Models: Lennard-Jones

- For a spherically symmetrical molecule, the classical second virial coefficient B depends on the potential energy, u, between pairs of molecules which is a function of separation distance between the molecules, r.

$$u(r) = 4\varepsilon\left[\left(\frac{\sigma}{r}\right)^{12} - \left(\frac{\sigma}{r}\right)^{6}\right]$$

$$B_{\text{classical}} = -2\pi\int_0^\infty\left[e^{-u(r)/kT} - 1\right]r^2 dr$$

The Lennard-Jones potential model

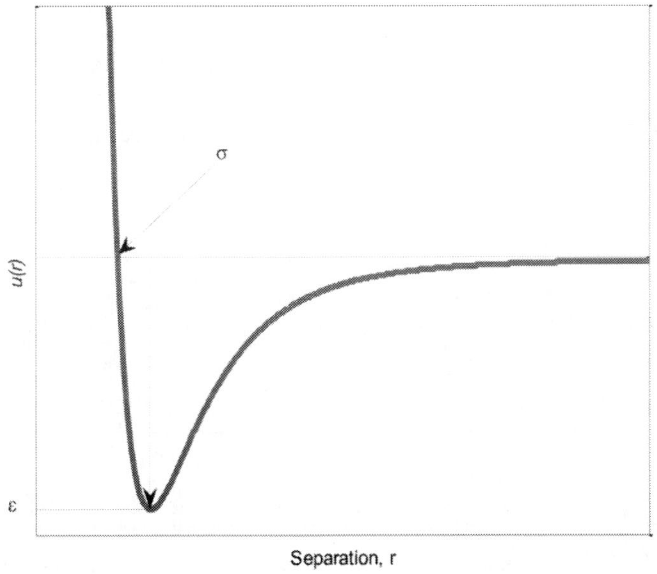

Potential energy, u, of two atoms as a function of separation, r

Potential Parameters

- An agreed upon unique set of collision diameters and well depths does not exist. This might be due to many factors including, but not limited to:

  1- The choice of experimental data, whether to use transport properties or thermophysical properties to obtain the potential parameters.

  2- Even with the same approach, some authors are inclined to use a certain data set. In the case of second virial coefficient, many authors prefer to use the data set of Malayshev which is very close to Morizot. However, others opt to use the data set of Heintz!

  3- What make things even worse are the different potential models used to obtain the parameters which renders comparisons cumbersome or meaningless.

Some Values of the Potential Parameters

Potential	Well depth (K)	Collision diameter (Å)	n	Ref
Lennard-Jones (6,n)	224	8.3	12	[46]
	258	7.1	12	[47]
	430	5.24	12	[35]
	399.8	6.5	12	[48]
	617.1	5.3	13.84	[34]
Kihara	1140	1.8	12	cited in [45]
Modified Buckingham (exp-6)	582	5.44	--	cited in [45]
Aziz MMSV	764	5.01	--	[45]
TDIP Zarkova	1044	4.992	27.87	[27]

How Potential Parameters are obtained?

- The second virial coefficient data was fitted to LJ potential using TDIP and TIIP by minimizing the Root Mean Square Deviation (RMSD) between experimental and calculated values.

$$\text{RMSD} = \sqrt{\frac{1}{O}\sum_{i=1}^{O}\left(B_{i,\exp} - B_{i,\text{cal}}\right)^2}$$

Zarkova & Hohm

$$u_{ij} = 4\varepsilon_{ij}(T)\left[\left(\frac{\sigma_{ij}(T)}{r_{ij}}\right)^{12} - \left(\frac{\sigma_{ij}(T)}{r_{ij}}\right)^{6}\right]$$

$$\sigma_{ij}(T) = \sigma_{ij}^{0} - \frac{\delta}{T}$$

$$\varepsilon_{ij}(T) = \varepsilon_{ij}^{0}\left(\frac{\sigma_{ij}^{0}}{\sigma_{ij}(T)}\right)^{6} = \varepsilon_{ij}^{\infty}\left(\frac{\sigma_{ij}^{0}}{\sigma_{ij}^{0} - \frac{\delta}{T}}\right)^{6}$$

Gibbs Ensemble Monte Carlo (GEMC)

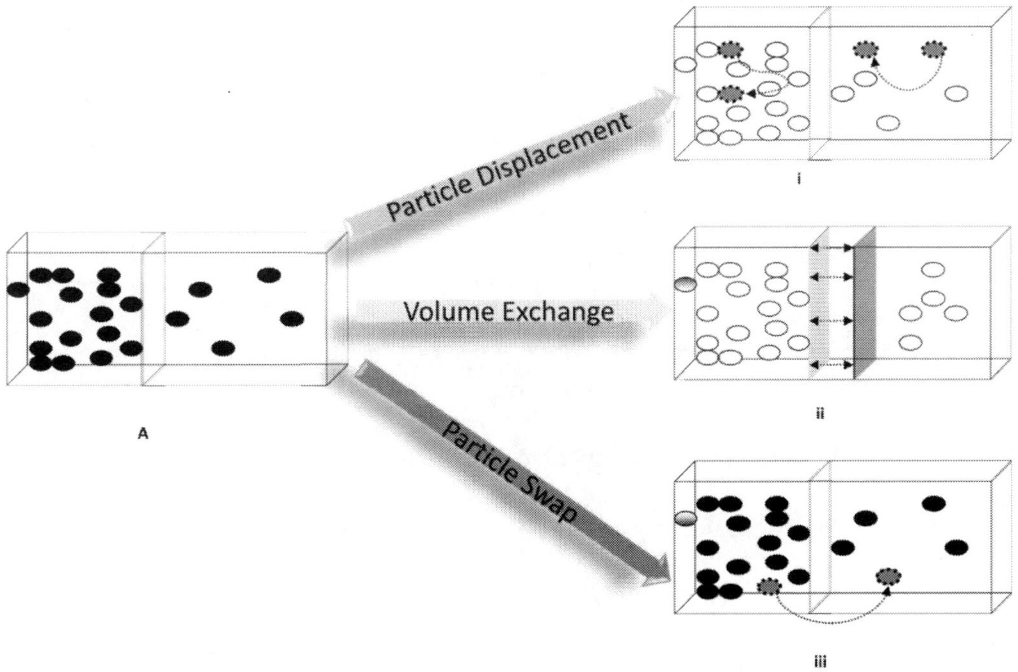

GEMC Simulation Details

- Gibbs ensemble was used to simulate the coexistence of vapor and liquid phases. Systems of size 1000 atoms were studied to simulate the vapor and liquid phases.
- The total number of particles was equally partitioned initially between two boxes.
- Initial configurations were generated on a simple cubic lattice.
- The temperature of the entire system is held constant and surface effects are avoided by applying periodic boundary condition.

GEMC Simulation Details

Property	
Number of molecules	2×500
Cutoff radius	$2.5\ \sigma$
Configurations generated	20,000,000
Particle displacement attempt	5000
Volume fluctuation attempt	1
Particle interchange attempt	1000
Equilibration cycles	1000
Production cycles	5000
Typical run time (hr.)	4

Results: TDIP and TIIP from Second Virial Coefficient Fitting

TIIP				TDIP			
σ (10^{10}Å)	δ(10^{10}Å.K)	ε/k_B (K)	RMSD (cm^3/mol)	σ(10^{10}Å)	δ(10^{10}Å.K)	ε/k_B (K)	RMSD (cm^3/mol)
8.126	0.000	236	11.2	5.887	99.45	324	3.9

used data set C

SVC Predictions for TIIP and TDIP

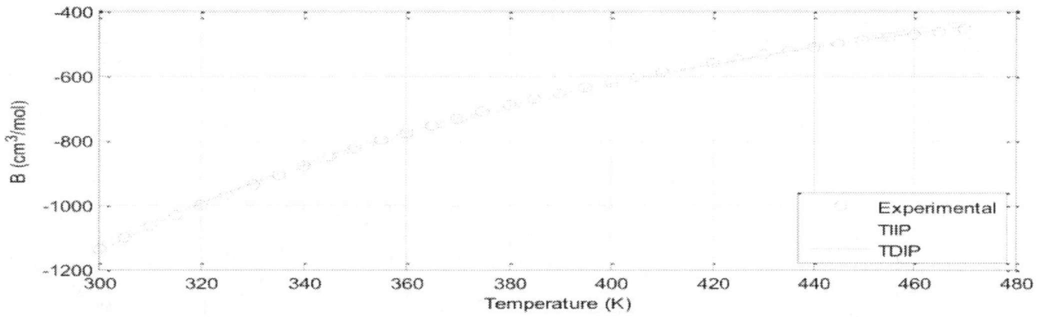

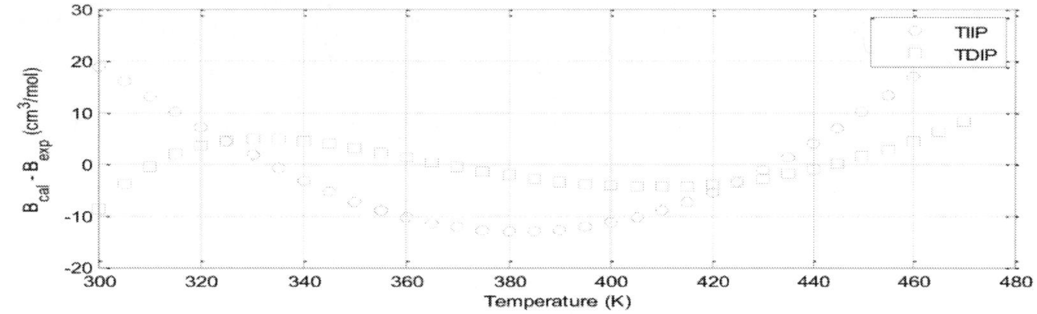

Comparison of our fitted values with previously published data

Potential	Well depth (K)	Collision diameter (Å)	n	Ref.	RMSD (cm^3/mol)
Lennard-Jones (6,12)	224	8.3	12	[46]	38.895
	258	7.1	12	[47]	151.31
	430	5.24	12	[35]	147.66
	399.8	6.5	12	[48]	261.01
TDIP Zarkova	1044	4.992	27.87	[27]	35.449
TIIP this work	236.13	8.1258	12		11.187
TDIP this work	323.63	5.8871 - 99.445/T	12		3.8931

GEMC Simulation Results: VLE Phase Diagram

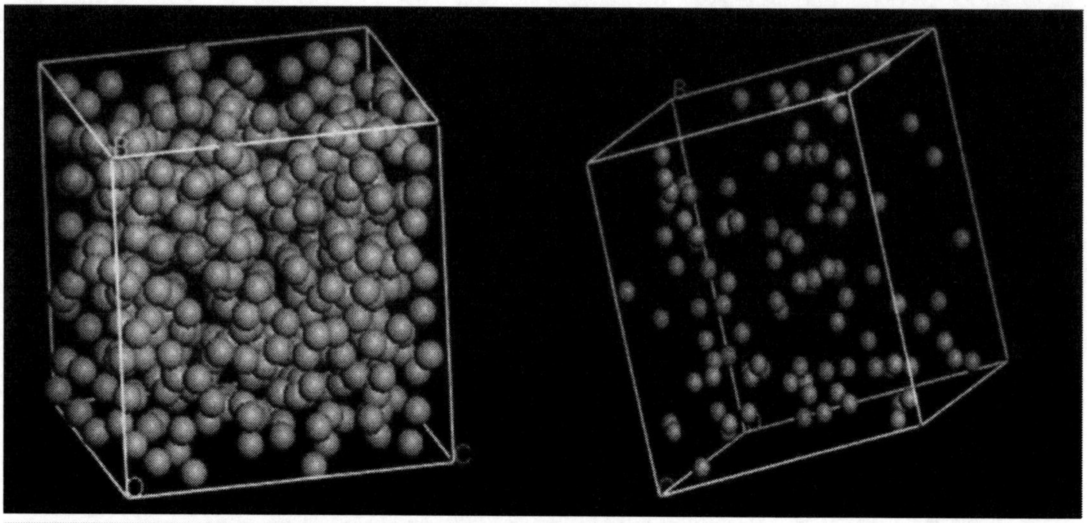

Typical snapshot of final configurations of UF_6 system of N=500 atoms at T=340 K. **(boxes not shown to scale)**

The left box is the liquid phase configuration.
The right box is the vapor phase configuration.

Quantitative Analysis of GEMC Simulation Results

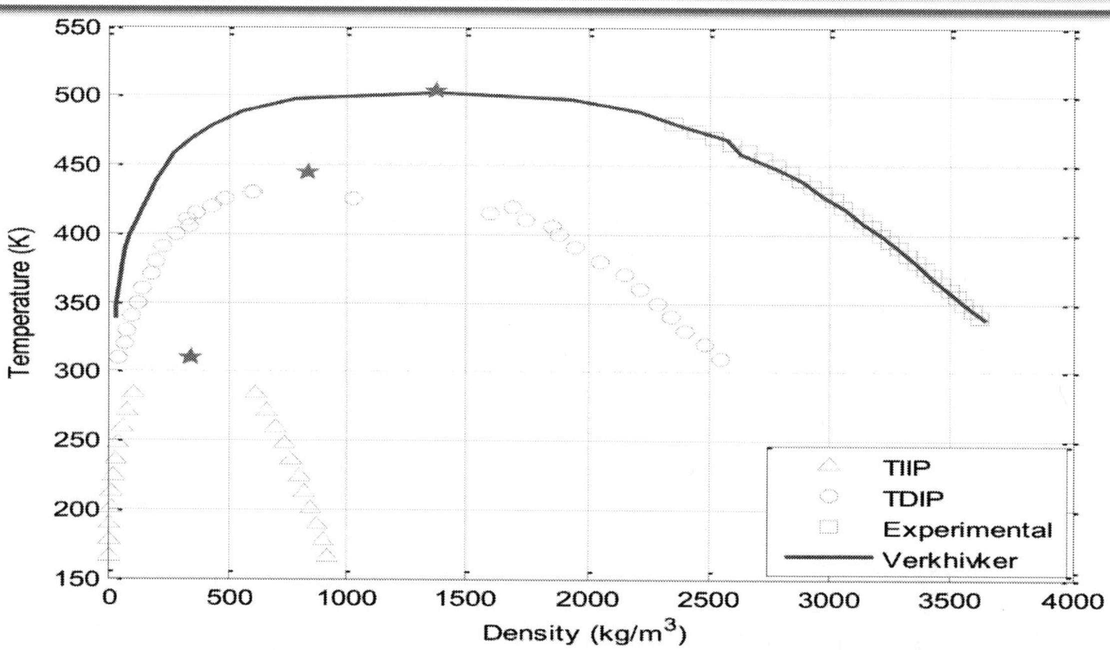

VLE phase diagram for UF_6 using TIIP, TDIP compared against experimental data

Critical Constants

	ρ_c (kg/m³)	P_c (bar)	T_c (K)
Experimental (calculated by extrapolation to the critical region [4, 53])	1375	46.1	503.3
TIIP (this work)	344.27	7.77	309.8
TDIP (this work)	831.95	31.1	444.3

Vapor Pressure vs. Temperature

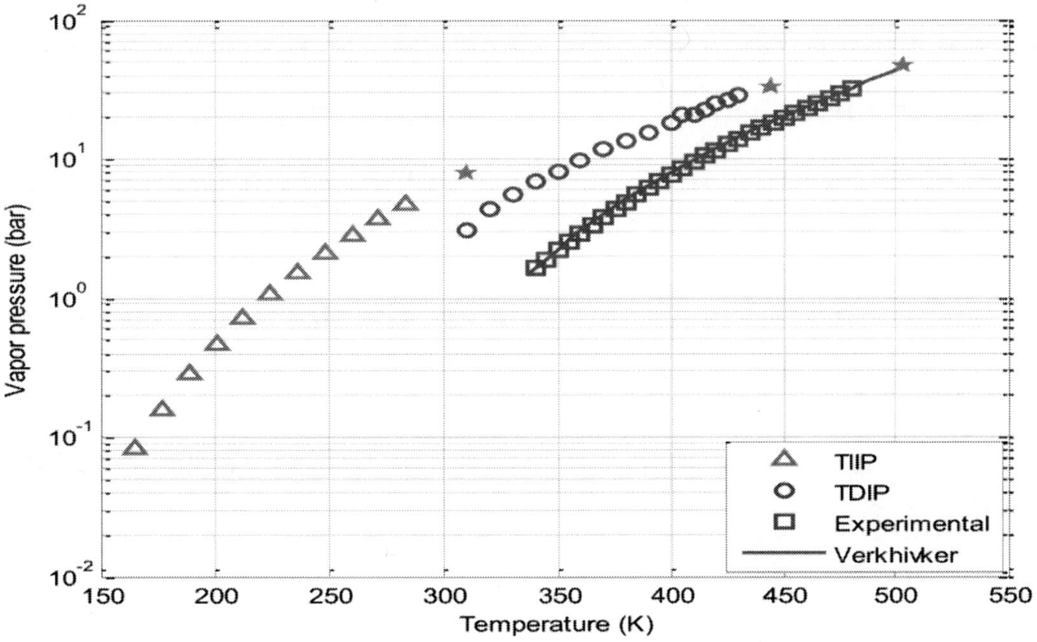

Conclusions

- Temperature dependent interaction parameters (TDIP) for UF_6 modeled as a united atom Lennard-Jones fluid is carried out. The collision diameter, $\sigma(T)$, and well depth, $\varepsilon(T)$ were obtained by fitting experimental second virial coefficient data.
- This reduces the RMSD by about 60% compared to conventional temperature independent interaction parameters (TIIP).
- The RMSD values are within the expected uncertainty of experimental data for the TDIP.

Conclusions

- We found that the TDIP improves the simulated vapor-liquid coexistence curve, critical constants, and vapor pressure of UF_6 modeled as a united atom Lennard-Jones fluid
- It can potentially improve prediction of any phase diagram.
- This finding is consistent with the previous work of the researchers on simple fluids.

Conclusions

- Here GEMC simulations using TDIP generated a phase diagram closer to its experimental counterpart. However, both TDIP and TIIP approaches need improvements since the predictions are still far from the experimental saturated liquid density data for UF_6.

Conclusions

- The shortcomings of the TDIP approach may be due to all, one or a combination of three plausible explanations:

  1- inadequacy of the united atom approach,

  2- inadequacy of the L-J model and its associated parameters,

  3- exclusion of Columbic electrostatic interactions.

Recommendations

- Accurate and consistent experimental data needs to be available for UF_6 to be able to generate reliable conclusions based on the parameters fitted.

- Fitting TDIP and TIIP using available liquid and vapor phase transport properties (self-diffusion coefficient and viscosity) instead of the second virial coefficient data.

- Columbic interactions were ignored in this work. Taking this work a further step and simulating the VLE including electrostatic interactions might improve the accuracy.

Recommendations

- Changing the LJ potential to more accurate form, though computationally more expensive, such as exp-6 or Kihara or fitting the repulsive exponent.

- Using TDIP in predicting transport properties of UF_6 via molecular dynamics (MD) simulations.

- Carrying out GEMC or MD simulations for multicomponent systems of interest is essential given the safety issues encountered in experimental measurements. Systems such as HF-UF_6 and solvent-UF_6 would be a first step in this direction.

Acknowledgments

- This work is carried out through generous funding from the University of Jordan (JU) during the granted sabbatical leave in the academic year 2014/2015 spent at King Fahd University of Petroleum & Minerals (KFUPM), Kingdom of Saudi Arabia.
- We would like to acknowledge the help and support furnished by KFUPM through Grant IN131040.

شكرا لحسن ألاستماع

36

498504 Prediction of Cesium Loading on Crystalline Silicotitanate Ion Exchange Resin

Abstract:

At the Savannah River Site (SRS) waste saltcake stored in the SRS Tank Farms is dissolved in water and processed through the Tank Closure Cesium Removal (TCCR) system to remove radioactive cesium and strontium from the high-activity waste solution. The TCCR system is a crystalline silicotitanate (CST) ion exchange (IX) system comprised of a carousel of two small lead-lag IX columns. Concerns have been raised that cesium could accumulate on the CST resin at concentrations that exceed the safety threshold of 170 Ci/L, potentially necessitating a costly redesign of the TCCR system.

A computer model was developed by Professor Rayford G. Anthony of Texas A&M University (TAMU) to simulate IX equilibria between electrolytic solutions and CST solid resin. The TAMU model was used to estimate cesium loading on CST as a function of initial cesium concentration in the waste solution.

This paper presents the computational results for a wide range of waste solution and discusses the effects of key ions (i.e., Na^+, OH^-, K^+) on cesium loading.

Computational Optimization of Difficult Physics-Based Project Designs Using Massively Parallel Solution Methods

2017 AAEES Excellence in Environmental Engineering and Science ™ Grand Prize Competition Winner - Research Category

Presenter:
Larry M. Deschaine, Ph.D., P.E.
AICHE Annual Meeting
November 1, 2017

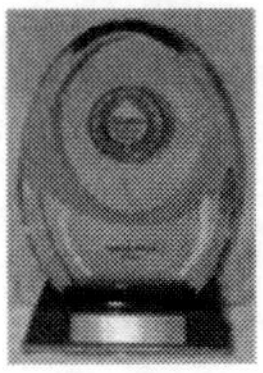

Why PBMO™ Subject Matter Expert Decision Support

- National Research Council (2013) findings
  - Large number of complex groundwater sites requiring cost effective long term management
- Distributed computing platform
  - Rapid & automated analysis of large number of simulations
- Link optimization algorithms with physics based model
  - Optimal pump and treat (today)
  - Source delineation
  - Plume delineation/Long term management
  - Treatment trains

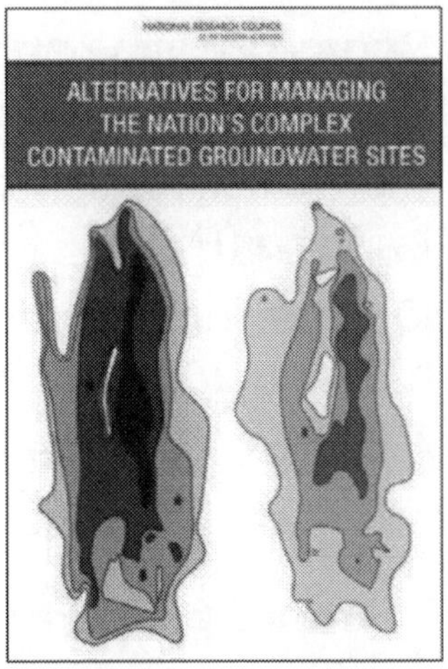

How can we bridge from where we are, to where we need to go?

2

Current Problems/Challenges

- **Time and Uncertainty:**
  - Pump-and-treat (P&T) remedies have been operating continuously for many years with little progress toward remedial goals
  - Most P&T remedies lack exit strategies
- **Cost:**
  - Hundreds of millions of dollars are being spent annually on P&T
  - Emerging contaminants such as the group of chemicals commonly referred to as "PFAS" (i.e., per- and polyfluoroalkyl substances), which are associated with foams used in firefighting, will lead to more extensive use of P&T remedies with increased costs.
- **Deficiency:**
  - Traditional approaches for remedial optimization are inefficient:
    - Expert opinion with little quantitative foundation
    - Trial-and-error "Brute Force" model simulations, limited number or over-simplified model simulations to achieve computational efficiency

HGL

3

Design Optimization Goals

Site Investigation

- Source location and identification (SLI)
- Plume delineation and tracking (PDT)
- Optimal long–term monitoring (LTMO)
- Minimize life-cycle cost

Remedial Optimization

- Control plume migration
- Optimize groundwater drawdown
- Minimize or cap cost
- Minimize or cap time
- Maximize or surpass contaminant removal or destruction goal

Honor property access restrictions and engineering constraints

HGL

4

Physics-Based Management Optimization ™ (PBMO ™)

▼ Cloud-based PBMO™ Tool:

- Uses *computational optimization algorithms* for improving cost, time and certainty of remedial projects
- Solves highly complex optimization problems *rapidly* (minutes to hours)
- Provides optimal solutions for *remedy design, implementation, maintenance and exit strategies*

▼ PBMO™:

- Accommodates public-domain MODFLOW/MT3DMS, and HGL's MODFLOW-SURFACT™ cutting-edge software for highly robust and efficient parallel simulations on the Cloud.

5

PBMO™ Medallion Conceptualization

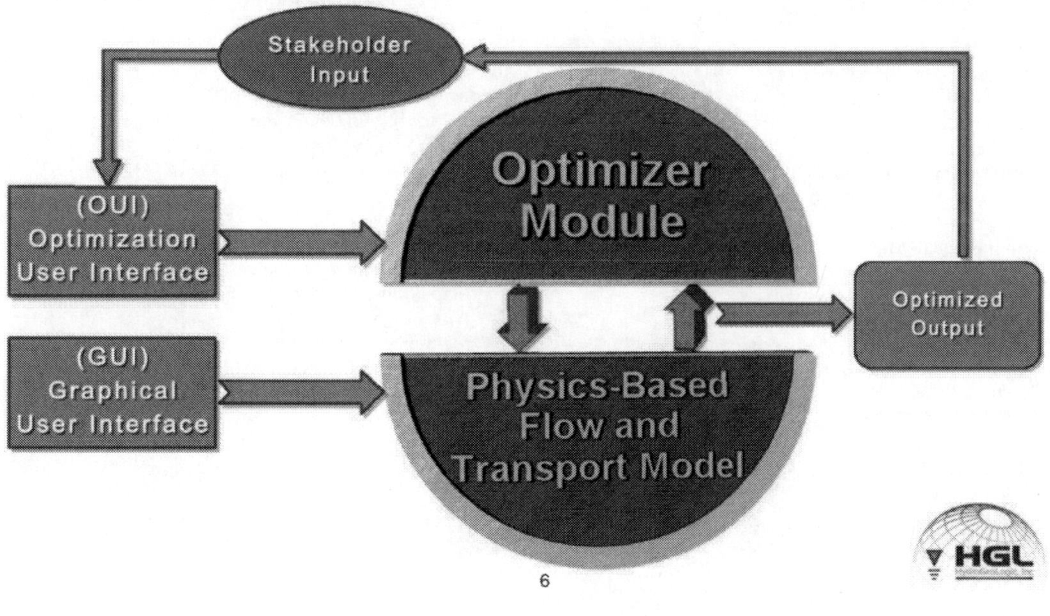

6

PBMO™ Optimization Tools

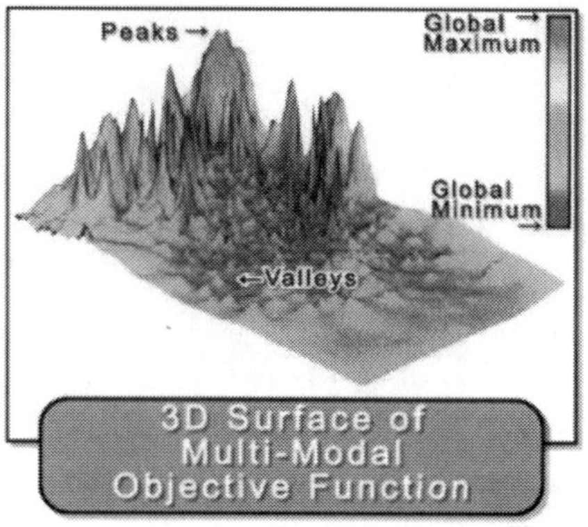

- Key elements of computational optimization include:
  - Objective Functions
  - Decision Variables
  - Constraints

Optimizer Module

PBMO™ Optimization Algorithms

Single Objective Deterministic Guided Search Algorithms:

Sequential Linear Approximation (SLA)
Sequential Quadratic Programming (SQP)
Generalized Reduced Gradient (GRG)
Branch and Bound (B&B)
Multi-Start with local search (MS+LS)
Lipschitz Global Optimization (LGO ©)
Constraint Programming (Logic-based optimizer)
-Semi-definite programming (SDP)
-Interior-point method (IPM)
-Multi-level coordinate search (MLCS)

Multiple Objective Algorithm:

Analytical Hierarchy Process (AHP)

Built with modern Fortran and Python

Parallel Search Methods

Global Adaptive Random Search (GARS)
Adaptive Global Random Search (AGRS)
Adaptive Design of Experiment (ADOE)
Monte-Carlo / Latin Hypercube
Multi-start Adaptive Partition (MSAP)

Machine Learning/Data Mining Algorithms:

WEKA (Machine Learning)
RapidMiner (Machine Learning)
Fuzzy Logic
Reinforcement Learning
Information Theoretics

Optimizer Module

Physics-Based Models in PBMO™

- ▼ MODFLOW-based groundwater flow and contaminant transport simulation codes
  - MODFLOW/MT3DMS
  - HGL's MODFLOW-SURFACT™
  - Future: MODFLOW-USG (unstructured gridding)
- ▼ Can be customized to accept codes for surface water and integrated surface water/groundwater modeling
  - HEC-RAS
  - HEC-HMS
  - MODHMS®

HGL

PBMO™ Execution on the Cloud

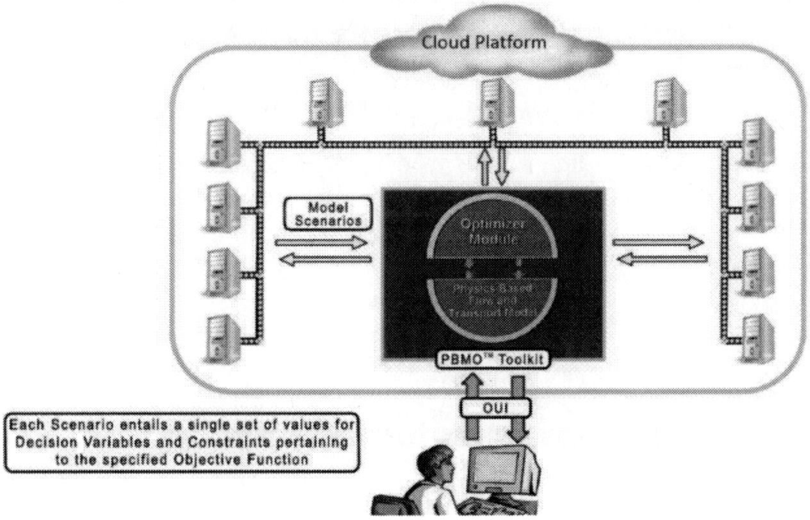

HGL

PBMO™ Optimization Computational Engine (OCE)

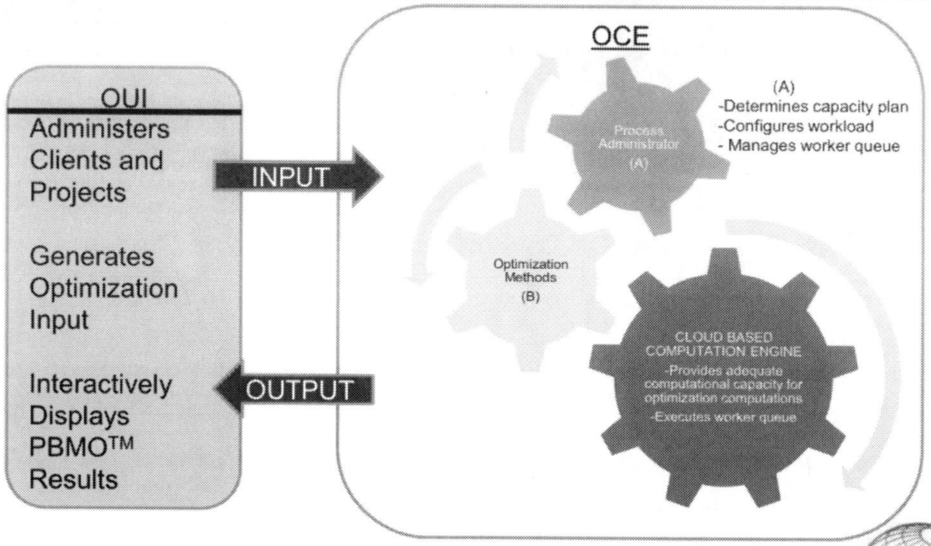

11

PBMO™ Optimization Computational Engine (OCE) – In Action

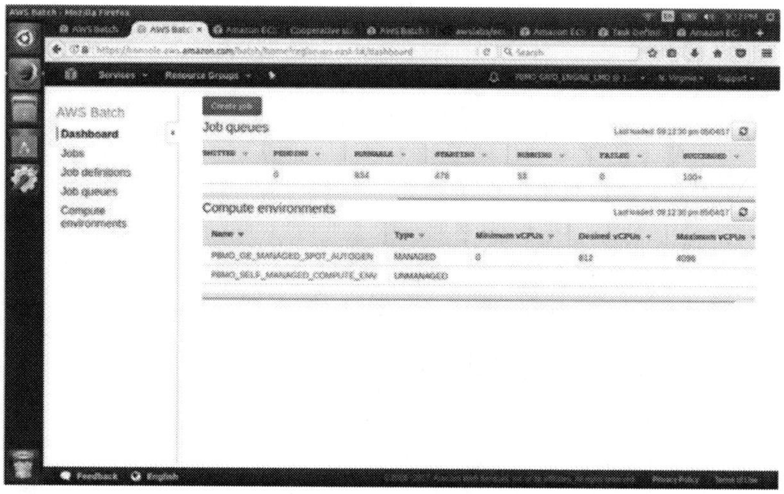

12

Example: Garvey Elevator Superfund Site, Hastings, NE
Site Location and Operational History

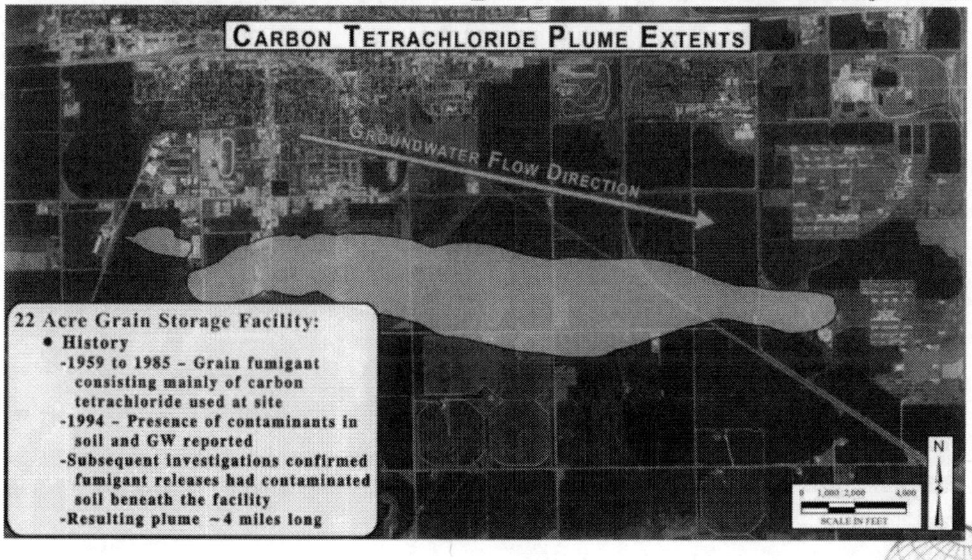

HGL

13

Example: Garvey Elevator Superfund Site
PBMO™ Design Criteria

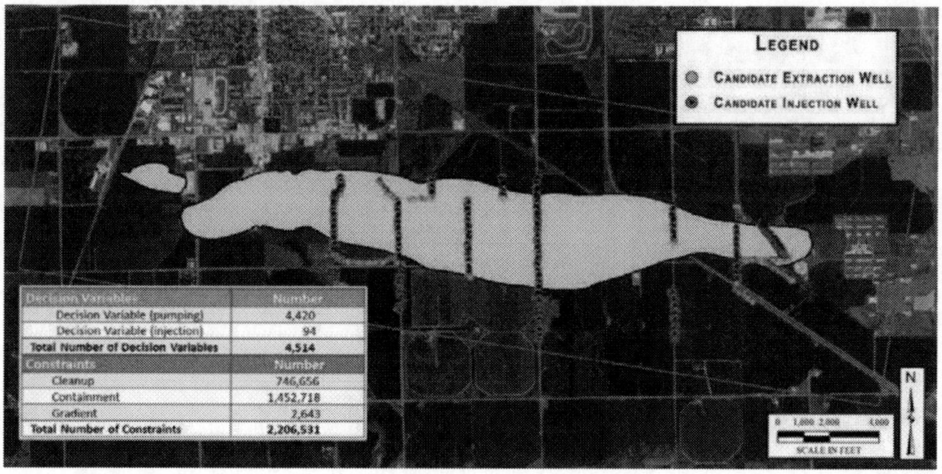

Decision Variables	Number
Decision Variable (pumping)	4,420
Decision Variable (injection)	94
Total Number of Decision Variables	**4,514**
Constraints	Number
Cleanup	746,656
Containment	1,452,718
Gradient	2,643
Total Number of Constraints	**2,206,531**

HGL

14

Example: Garvey Elevator Superfund Site PBMO™ Optimal Solution

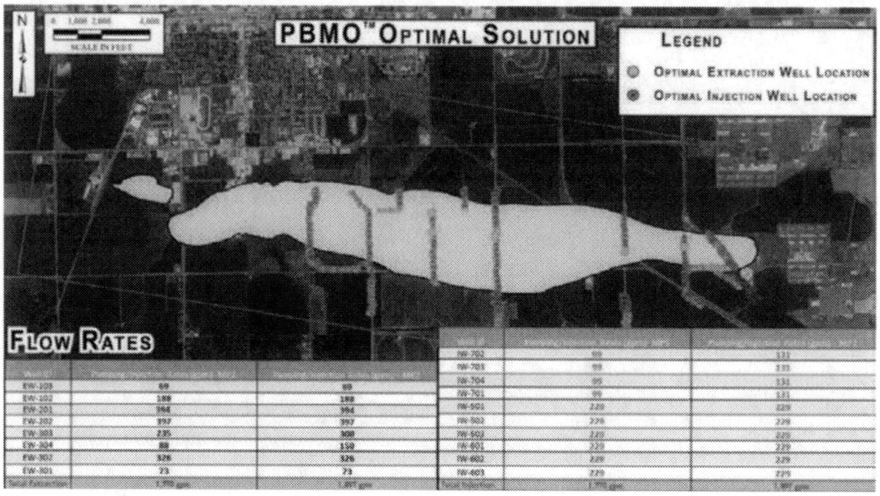

Example: Garvey Elevator Superfund Site Remedy Performance Snapshots

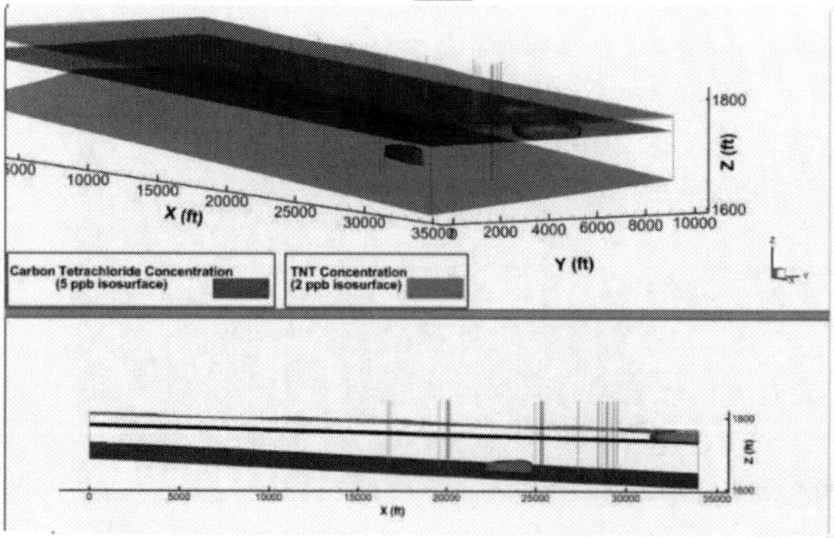

PBMO™ Future Enhancements

Problems and Challenges

- ▼ Optimal management of complex groundwater restoration sites
  - Limited characterization data
  - Multiple plausible conceptualizations
  - Incorporating treatment trains
  - Changing regulatory framework
  - Stakeholder acceptance

PBMO™ Future Enhancements

- ▼ Source location identification (SLI)
- ▼ Plume delineation and tracking (PDT)
- ▼ Optimal long–term monitoring (LTMO)

17

PBMO™ Future Enhancements: Source Location Identification (SLI)

- ▼ Optimal identification of source location is crucial for determining the nature and extent of groundwater contamination
- ▼ Sources consist of DNAPL, LNAPL and emerging contaminants (PFAS/PFOA)
- ▼ Sources of contaminants can be emanating from an individual source or from multiple sources

18

Source Delineation Concept

- Begin with Multiple Potential Source Locations with Various Likelihoods of being Actual

Subject matter experts assess potential source locations, and likelihoods

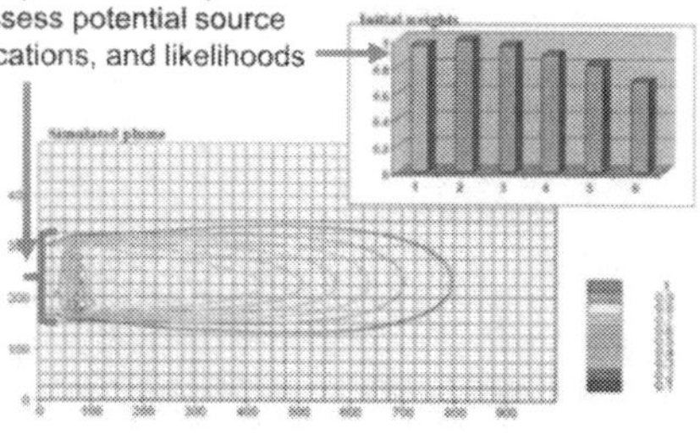

- End with the Source Location(s) Identified

Algorithm automatically develops sampling strategy to identify the true source area

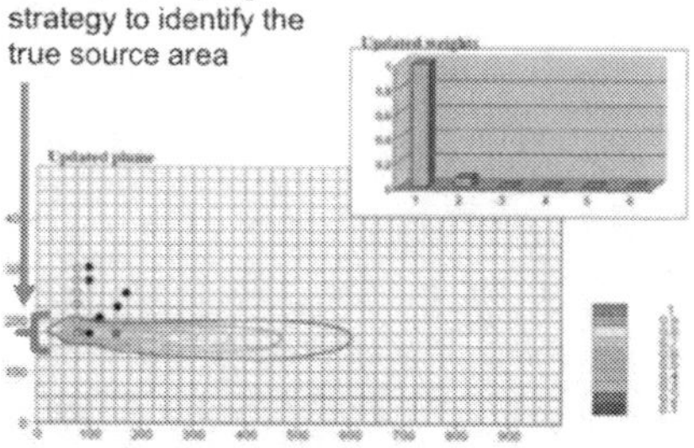

Iterative Sampling locations (●)

Ref: Pinder, G. F., Ross, J., Dokou, Z. Optimal Search Strategy of the Definition of DNAPL Source Location, SERDP, 2009

HGL

Source Delineation Site Demonstration

- Demonstration Report

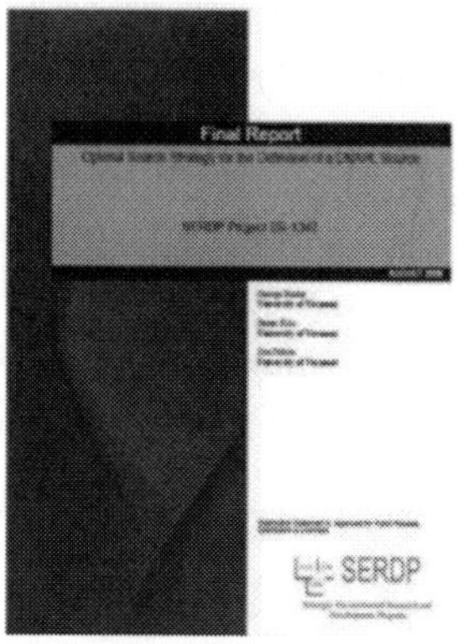

- Technology demonstrated at the Anniston Army Depot in Anniston, AL
- Multi-layer model (residuum overlying weathered bedrock)

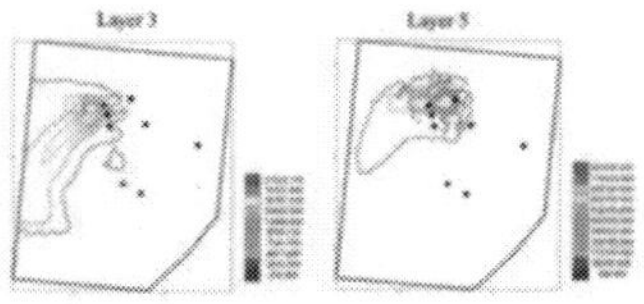

Results at Final Iteration

HGL is extending this technology to incorporate it into PBMO™

HGL

PBMO™ Future Enhancements: Plume Delineation and Tracking (PDT)

- ▼ Enables stakeholders and decision makers to base plume migration mitigation actions on a solid informational foundation
- ▼ Tools being incorporated in PBMO™ include:
  - Optimal monitoring well network design
    - ➢ Locate monitoring wells to maximize information content at a specified point in time
  - Long-term monitoring optimization
    - ➢ Locate monitoring wells to maximize information content at a specified time horizon

21

Case Study: TCE Monitoring Well Assessment for the Ogallala Aquifer

- ▼ Stakeholder Reviews:
  - Independent Review Team Recommends Formation of Technical Advisory Group (TAG)
  - TAG Unanimously Recommends Customized Tool for Plant Evaluation (Risk Assessment & Corrective Measures Study)
  - Application of Plume Delineation Module Initiated
- ▼ Results: The Plume Delineation Technology Assessed That:
  - The Existing MW Network Reduced Uncertainty by 93.4%. Optimal by 98%.
  - Only 1-2 Additional Optimally Located Wells Required Versus the Initial 6-12 Deep (~600 Feet) Wells Envisioned by the Stakeholders
- ▼ Acceptance: All Analyses and Conclusions were Accepted by Stakeholders
  - High Confidence in Analyses
- ▼ ROI Cost Savings:
  - Cost of Analysis: $96K
  - Well Avoidance Costs: $2M
  - ROI: 20:1

22

Presentation Summary and Conclusions

- New paradigm for decision support pertaining to environmental restoration and water resource management
  - Management optimization linked with physics-based models on distributed computing platform to provide real-time solutions
- Proven applications on P&T systems
  - Optimization of existing systems
  - Optimal design for new systems
- Expanding capabilities to provide decision support on all aspects of large, complex environmental restoration and water resource management challenges

23

Contact for Additional Information

Larry M. Deschaine, Ph.D., P.E.

HydroGeoLogic, Inc.

ldeschaine@hgl.com

24

504253 An Evaluation of Organically Bound Tritium (OBT) Research

Abstract:

Tritium (^3H) is radioactive isotope of hydrogen which decays via beta emission. Tritium incorporated into the tissues of plants and animals is referred to as organically bound tritium (OBT). Tritium ingested as tritiated water or water vapor (HTO) provides a limited dose that is subject to rapid turnover in animals. OBT, largely formed through photosynthesis, represents a more sustained dose when ingested. Elevated radiation exposure to DNA in cells can cause mutation that alter cell function and a serious may result, e.g., cancer. That is, just as water containing ordinary hydrogen and oxygen is a component of all living cells, HTO can also be incorporated into the cells of the body. Recent evidence demonstrates incorporation of OBT into plant cells so consumption by animals through the food chain is also possible. Research indicates that a fraction of ingested tritium can remain in the human body for more than ten years. However, dose standards for OBT do not exist and the literature highlights this gap.

The objective of this study is to assess organically bound tritium (OBT) research with a focus on impacts on organisms, including plants and animals. A thorough literature review of OBT was performed using various databases and reports from federal agencies. Research findings offer measures and influences of OBT however, standard protocols for assessing OBT is suggested.

489606 3D-Interconnected PVA-Citric Acid Porous Hydrogels Embedded with Potassium Copper Hexacyanoferrate Nanoparticles for Enhanced Removal of Radionuclide Cesium

Abstract:

A potassium copper hexacyanoferrate-embedded poly(vinyl alcohol)-citric acid hydrogel film (HPC) was prepared via a facile two-step method for the effective Cs^+ removal. The synthetic procedure consisted of the simultaneous Cu dispersion and incorporation, followed by the diffusion of potassium hexacyanoferrate accelerated by acetone evaporation. The diffusion-derived KCuHCF formation in the preformed hydrogel facilitated not only the preservation of the 3D-interconnected hydrogel structure, but high dispersion of the nanostructured KCuHCF. Using acetone as a non-solvent, the reverse diffusion rate of the incorporated Cu in the hydrogel matrix was greatly reduced; hence a large amount of KCuHCF was loaded in the matrix.

The as-obtained HPC demonstrates the superior Cs^+ adsorption capacity and selectivity with significantly fast removal kinetics. In the adsorption isotherm, the HPC showed a very high Cs^+ uptake of 667 mg/g KCuHCF at the saturated concentration. Moreover, the uptake process was exceptionally fast with 99.5 % of Cs^+ ions adsorbed within 30 min, and the overall Cs^+ removal was 99.9 % from a dilute Cs^+ solution (9.18 ppm). Even when tested in seawater, the HPC exhibited almost unaltered Cs^+ removal efficiency above 99.5 %, and the high distribution coefficient K_d value of 7.7×10^5 mL/g at an extremely low Cs^+ concentration (0.67 ppm, V/m = 1000 ml/g), which highlighted the exceptional Cs^+ affinity over all competing cations. In addition, this novel adsorbent represented the stable and high Cs^+ removal efficiency of 99.9 % over a broad pH range from 2 to 10.

These enhanced removal properties are closely related to the synergistic effects of rich K^+ contents in KCuHCF, well dispersion of the nanoparticles, and the fast swelling kinetics of the hydrogel matrix. Therefore, combined with the simple synthetic procedure and low production cost, the HPC is one of the most promising adsorbents for the rapid and selective removal of Cs^+ from contaminated environments.

AUTHOR INDEX